T0288136

Nearly Nuclear

Nearly

Nuclear

A MISMANAGED ENERGY TRANSITION

LeRoy Smith

MICHIGAN STATE UNIVERSITY PRESS | *East Lansing*

♾ The paper used in this publication meets the minimum requirements
of ANSI/NISO Z39.48-1992 (R 1997) (Permanence of Paper).

Michigan State University Press
East Lansing, Michigan 48823-5245

Library of Congress Cataloging-in-Publication Data is available
ISBN 978-1-61186-398-7 (paper)
ISBN 978-1-60917-672-3 (PDF)
ISBN 978-1-62895-440-1 (ePub)
ISBN 978-1-62896-434-9 (Kindle)

Book design by Charlie Sharp, Sharp Des!gns, East Lansing, MI
Cover design by Erin Kirk
COVER ART: Consumers Power Company, "Vicinity Map," n.d. [late 1960s],
Folder: Consumers Power Documents 1969–1970, Box 4,
Mary P. Sinclair Papers, Bentley Historical Library, University of Michigan.

Michigan State University Press is a member of the Green Press Initiative and is committed to developing and encouraging ecologically responsible publishing practices. For more information about the Green Press Initiative and the use of recycled paper in book publishing, please visit *www.greenpressinitiative.org.*

Visit Michigan State University Press at *www.msupress.org*

Dedicated to my wife, H.J.
She was always present with encouraging words
as I labored to finish this book.

Contents

ix PREFACE

xiii ACKNOWLEDGMENTS

1 CHAPTER 1. Location, Location, Location: Geology Is Destiny

7 CHAPTER 2. Imagine This: A Whole Town Rallies in Support of a Nuclear Power Plant

13 CHAPTER 3. Regulators: Safety Rules and the Spanish Inquisition

19 CHAPTER 4. To Build a Nuclear Power Plant: Admiral Rickover's Way

27 CHAPTER 5. A Utility and a Chemical Company: Can This Marriage Endure?

35 CHAPTER 6. Consumers Power: Challenges of Managing New Notions

45 CHAPTER 7. Let Bechtel Do It: Just Make It a Megaproject

49 CHAPTER 8. 1968: Forces Not under Your Control Matter

55 CHAPTER 9. Mary Sinclair: No Typical 1960s Woman

65 CHAPTER 10. How to React to Midland Plant Critics: Shame, Organize, Label

73 CHAPTER 11. Midland Pride: Why Midlanders Supported the Midland Nuclear Plant

79 CHAPTER 12. Myron Cherry: Bright, Dedicated, and Mean

89 CHAPTER 13. Construction Quality Control: An Unredeemed Promise of Reformation

95 CHAPTER 14. 1975: Carry on with the Midland Plant

103 CHAPTER 15. Midland Plant: Opportunities (and Frustrations) for All

111 CHAPTER 16. 1978: Not Only Optimism Sinks and Cracks

121 CHAPTER 17. TMI2: Definitely Not Too Much Information

129 CHAPTER 18. People Power: Four Thousand March

137 CHAPTER 19. The 1980s: Problems Intensify As the Money Drains

147 CHAPTER 20. Quality Control Problems: Déjà Vu

157 CHAPTER 21. Brink of Bankruptcy: No Sales Contract, No Cost Recovery, and No Money

165 CHAPTER 22. 1985: What Happened Next?

171 CHAPTER 23. Consumers Power to Consumers Energy: From Nuclear to Natural Gas and Beyond

181 EPILOGUE

185 APPENDIX 1. Radioactive Waste Concerns

187 APPENDIX 2. Nuclear Reactor Types

189 APPENDIX 3. Simplified Diagram of the Midland Plant

191 APPENDIX 4. Fix for Cracking in the Walls of the Auxiliary Building

195 APPENDIX 5. Conversion of the Midland Plant to a Natural Gas Plant

199 NOTES

247 SELECTED BIBLIOGRAPHY

251 INDEX

Preface

This story is about Consumers Power Company's attempt—and ultimate failure—to build two huge nuclear reactors in Midland, Michigan. From its very announcement in 1967, this project was a continuing saga of apparent successes turning into failures and a failure eventually turning into a success. At one point the project was called an engineering nightmare, a financial disaster, a construction boondoggle, a political headache, and a regulatory muddle. Understanding why Consumers Power's initial, ambitious plan to provide clean, low-cost power failed is critical today if we are to manage the transition away from fossil fuels to renewables necessary to mitigate climate change.

Nearly Nuclear tells a story about the successes and failures of all types of people (construction workers, construction managers, environmental activists, businessmen, government regulators, lawyers, and politicians) and about how local, national, and international events catalyzed the reactions that drove this story. This nuclear power plant promised to make workers' jobs more secure and Midland's air cleaner by producing low-cost steam and electricity, but environmental activists thought the plant would not be safe. Construction workers were eager for the opportunity to ply their trades in building the plant, but government agencies imposed ever-changing regulations and design specifications. Not only was Consumers Power unable to provide construction oversight and financing to finish the plant, lawyers and politicians could not resolve the seemingly endless disputes that arose in connection with building it.

In 1984 after spending more than $4 billion ($13 billion in 2020 dollars) over a period of seventeen years, Consumers Power cancelled the nuclear plant. But this failure was not the end of the story. When a new partnership formed to use a portion of the equipment constructed for the nuclear plant for a natural gas-fired power plant, the original goal of providing a source of power without the soot and sulfur dioxide (acid rain) produced by coal-fired generation was accomplished. This natural gas-fired plant commenced commercial operation in 1990 and has now run successfully for over thirty years.

One could be tempted to say that the outcome of this story was inevitable. It was not. Although many other nuclear power plants were cancelled during the time of this story, about 100 were finished and went on to provide about 20 percent of the electricity in the United States between 1990 to 2020. The reasons why the Midland project failed are complex and shed light on the complicated work that will be required to shift our power generation to non–carbon-producing fuels.

Although this is a local history of Midland from the 1960s through the 1980s, I have recorded it because it provides a case study of the complex intersections of project and financial management, environmental activism, regulatory oversight, and politics. To understand any case history requires attention to the context in which it happens.

I must at this point give a disclaimer and a clarification. Although I spent a career dealing with natural gas, my telling of the *Nearly Nuclear* story is not an appeal for more power generation using natural gas. Although I am a certified petroleum geologist and a lifelong member of the American Association of Petroleum Geologists, I believe climate change is real, and the major contributor to it is the carbon dioxide that is dumped into the air by burning fossil fuels (including natural gas). Natural gas is a useful transition fuel for generating electricity because it is quickly available at any time when needed and only releases approximately half the carbon dioxide that coal does. Until such time as wind and solar power combined with the necessary electric storage and transmission grid are available for an all emission-free future, natural gas will play an important role in providing a reliable source of electricity. I do believe that as quickly as possible America should move to generation that adds no carbon dioxide to the air.

But in order to do so, those responsible for that transition will need to successfully navigate public opinion, regulatory, business management, and financial challenges. I hope that knowledge of the people and events central to this story will help those involved in implementing the new power grid in the 2020s to avoid some—if not all—of the difficulties encountered in Midland as they search for creative solutions.

Acknowledgments

I must first say that the book you are holding bears no resemblance to what I set out to write about the Midland Plant, and it can all be blamed on geology. My career had several parts. First I worked as a petroleum geologist and migrated over into the business of evaluating natural gas supplies for businesses that used natural gas for fuel. Then I managed gas supply and short-term electric sales for the Midland Cogeneration Venture Limited Partnership (MCV), which is located where an attempt had been made to construct the Midland Nuclear Plant.

Even though in the latter part of my career I was not involved in geology on a day-to-day basis, I retained a strong interest in all things geological and attended the meetings of the Michigan Basin Geological Society (MBGS). The January 2016 meeting featured a talk on using a seismic method in Michigan to determine the thickness of glacial deposits overlying bedrock. In the question and answer period a member in the audience asked if, in addition to determining the thickness of glacial deposits, this technique could also determine if deposits of peat were present besides the usual sand, clay, gravel, and glacially transported rocks. The questioner went on to say that if you could have been able to determine the presence of a peat bed under where the structures of the Midland Plant were built, the structures could have been located elsewhere, thus avoiding a very serious problem.

I was keenly interested in this discussion. During the sixteen years I had worked at the site of the Midland Nuclear Plant, I talked with many of the engineers who had worked on the original construction, and I had never heard any mention of a peat bed under the site as the cause of the structures' sinking and cracking. The questioner said I could obtain more information by talking to a geologist who had examined the cores recovered from drilling on the Midland site. My thought was that with a source who could verify the existence of the peat bed, I would write a technical article or perhaps just a short general interest article about it for the *Midland Daily News* or the magazine *Michigan History*.

After four months I finally located the geologist who was supposed to have first-hand knowledge of the peat bed. During this four month period, I looked at material on the Midland Plant and became fascinated with the complete history. When I contacted him, he told me that he didn't remember any peat bed. Now the problem was that by the time I discovered this, I was hooked on the story of the Midland Plant.

I had found in the past that when I get hooked on a subject, it is good to give a lecture on the topic and then I can let it go. So I would like to thank the Midland Noon Rotary Club; Midland's local television station, MCTV and Ron Beacom; the Midland Historical Society; the Osher Lifelong Learning Institute at Aquinas College; and the Saginaw Valley Torch Club who all graciously allowed me to give my lectures on the Midland Plant. But preparing these lectures only increased my obsession to determine what happened at the Midland Plant. So I continued to gather more material on the subject.

I need to thank many people for locating the material that hooked me on this story. Midland is blessed with the Herbert D. Doan Midland County History Center and the Midland County Historical Society. At the Doan Center, Crystal Laudeman was of great assistance in locating material. Michigan is fortunate to have the Bentley Historical Library at the University of Michigan where the papers of Mary Sinclair are available for researchers. I appreciate all of the assistance of that library's staff.

I am grateful for the academic and municipal libraries whose librarians assisted in my quest to find information. These institutions not only have books available, they also preserve newspapers, clippings, company documents, government documents, magazines, and journal articles. Librarians at the following libraries provided me with courteous assistance in locating documents for my research: Alice and Jack Wirt Public Library in Bay City, Jackson Carnegie Library in Jackson,

Charles V. Park Library at Central Michigan University, Delta College Library at Delta College, Hoyt Public Library in Saginaw, Walter Reuther Library in Detroit, and the Grace A. Dow Memorial Library in Midland. I would especially like to thank the librarians with the Grace A. Dow Memorial Library who had to deal with all of my numerous requests for old copies of the *Midland Daily News* as well as Stacy McNally at the Hoyt Library in Saginaw, Thomas Trombley at the Castle Saginaw County Museum, and Mary Wallace at the Walter P. Reuther Library who helped me locate pictures.

But as I continued to accumulate more and more material, my problem was what could I possibly do with it? Enter my wife, H. J., who spent a career teaching the writing process. This book could not have been written without her continuing encouragement and guidance. Little did she know that, after a career of guiding and encouraging thousands of students to write, in retirement she would have to deal with her most challenged student, me. This book should have been titled *Nearly Nuclear: As Told to H. J. Smith.* In spite of all of her help I still slip into the passive voice, construct garbled sentences, and spell abominably.

A number of my friends were kind enough to read a few parts of my early manuscript, and they gave me encouragement. I thank Mike Cronenberger, Edward Rogers, Elyse Rogers, Bruce Peck, and Gary Skory for their efforts on my behalf.

But to call something you have written a book rather than a manuscript, one needs to find a publisher. The publishing of this book was made possible by Catherine Cocks, Michigan State University Press editor-in-chief, who saw in my rough manuscript something that could be molded into a readable book. Her guidance, support, and comments have made this book a reality.

Dave Shane and Kate Hessling of the *Midland Daily News* deserve special credit for giving me access to their photographic collection and allowing me to use their pictures in this book. I would especially like to thank Victoria Ritter of the *Midland Daily News* for assisting me in my picture quest, first in the attic of the old *Midland Daily News* building and then in the basement of their new office. An important data source I used was the annual reports of the Consumers Power Company. I thank Consumers Energy Corporate Library and Thomas Begin for their generous contribution to this project by supplying me with copies of annual reports for the Consumers Power Company for the years 1967 to 1984. I also thank the MCV and Doyle Beneby for permission to publish a picture of their facility.

Last but not least, I wish to thank the people who gave me their time to tell me what they remembered about the Consumers Power Company, the Dow Chemical

Company, the city of Midland, the Midland Plant, Michigan politics, and what followed the cancellation of the nuclear Midland Plant. Although there is a great deal of information available in print, microfilm, film, and electronic data on the story of the Midland Plant, input from real people was critical to this project. I would, however, emphasize that what you will read here is my interpretation of what I was told; I may not have completely understood what I was told and read or dug deep enough for sources on my subject. Not until I undertook this project did I understand the difficulties of recording history.

Although I may inadvertently leave someone out, I wish to thank the following people whose names can't be found in the text of this book, but who provided me with assistance and contributed to what came to be my understanding and my telling of the history I have written: Tom Adams, Howard Alexander, John Bartos, Lynne Bernabei, Naren Bhatia, Dan Bishop, Eric Blackhurst, Ruth Blair, Tom Bradley, Ned Brandt, Scott Brooks, Jack Brown, Kathy Brown, Steve Buchanan, John Catenacci, Marilyn Clark, Nina Cronenberger, Gabrielle Despaigne, Paul Duso, Deb Ege, Tom Elward, Samantha Engel, Doug Enos, Ronald Erickson, William Gentry, Gene Gillette, Mike Gllis, Jordan Howell, Preston Jones, John Kindinger, Jana Kullick, Bob Lewis, Mary Stinton Marshall, Francis Martin, John McMurray, Cliff Miles, Sharon Mortensen, Geoff Newhouse, Kevin Olling, Jim Rajewski, Andy Rapp, Dione Rush, Herb Scogg, Gary Smith, Bill Swenson, Laurie Valasek, Jenée Velasquez, Derk Vokal, Brian Wheeler, Mike Whiting, Anne Wolfe, Anastasia Wraight, Lowell Youngquist, and Rebecca Zeiss.

My final acknowledgment is to Hal Leonard LLC, from whom I received permission to use the lyrics from the *Mary Poppins* song, "Fidelity Fiduciary Bank," in this book.

CHAPTER 1

Location, Location, Location

Geology Is Destiny

Geology matters in the price of power, and in order to understand this story, a brief natural resources primer is helpful. In the 1700s maps of Michigan's Lower Peninsula showed a range of mountains trending north to south in its center.[1] However, these mountains were pure cartographic fiction. Indeed, rather than the heaving geologic forces that push huge plates together to form mountains, the rocks along a northwest to southeast corridor under the Lower Peninsula began sinking rather than rising over five hundred million years ago, in what geologists now call the Cambrian Period. For the next hundreds of millions of years, the area continued to subside, forming a large shallow bowl called the Michigan Basin.[2] Inland seas advanced and retreated into this bowl. The sediments these seas deposited were at times favorable to the formation and trapping of oil and gas deposits, but the quantity of that oil and gas would fall far short of Michigan's twentieth-century energy requirements.

From time to time the seawater evaporated from these shallow inland seas, leaving behind brine in small, interconnected voids in the sediments. Over time these sediments became rocks, and the brine in them contained a high concentration of salts.[3] At other times the seawater completely evaporated and left deposits of salts. In addition to common salt (sodium chloride), these brines and layers of

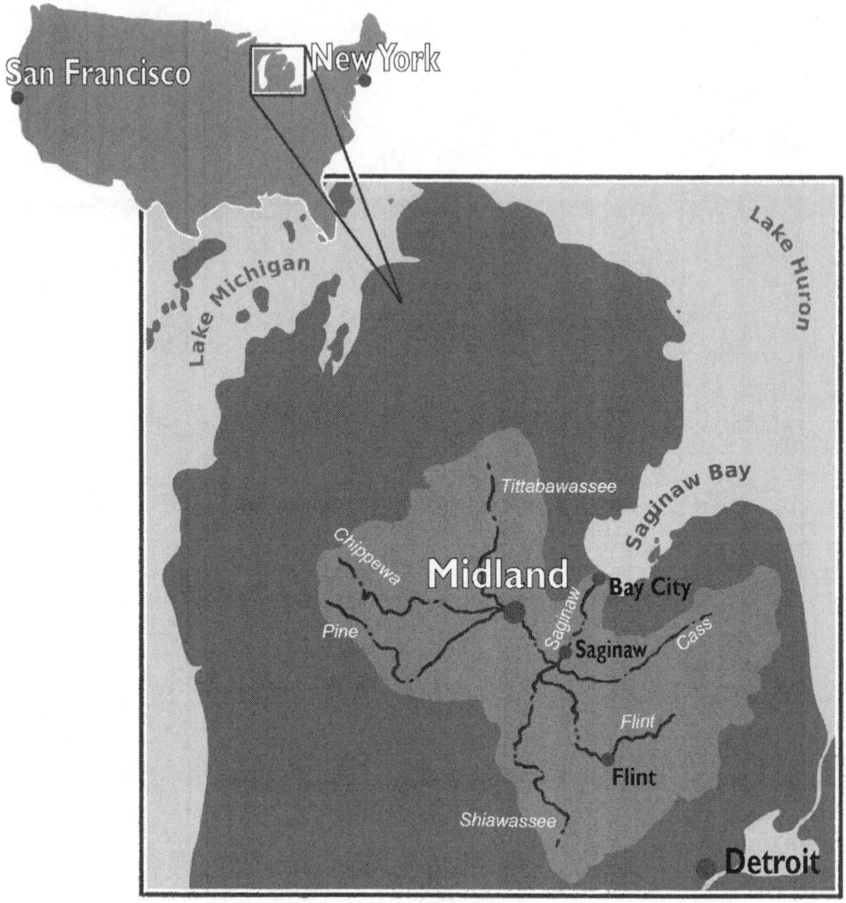

MAP 1. Midland is located on the Tittabawassee River in the Saginaw Valley (light gray area). Other cities in the Saginaw Valley with populations over thirty thousand are Bay City, Flint, and Saginaw. (This map includes a modification of a map of the Saginaw River watershed that can be found at https://en.wikipedia.org/wiki/Saginaw_River#/media/File:Saginawrivermap.png.)

salts included a broad class of compounds containing magnesium, calcium, and bromine.[4] In the nineteenth and twentieth centuries, the Dow Chemical Company would put these compounds to industrial use.[5]

During the Pennsylvanian subperiod about three hundred million years ago, when deposits of organic matter that would become thick deposits of coal were accumulating in other locations in North America, identical conditions existed in the central part of the Lower Peninsula. But in Michigan only discontinuous, thin coal beds (usually less than a few feet thick) formed.[6]

Millions of years after the rocks containing Michigan's limited supplies of coal, oil, and natural gas formed, glaciers advanced and retreated numerous times over the Lower Peninsula. In some places where the glaciers stopped advancing, they left hills of rocks and sand with elevations of only a few hundred feet.[7] This lack of significant elevation differences meant limited capacity for generating hydroelectricity by damming rivers—a significant issue in this story.

As the glaciers retreated, rivers and lakes formed, and clay was often deposited in the bottoms of these lakes. Lake Huron formed east of what is now Midland. Successive advances of lobes of ice scoured out the Saginaw Bay, which filled with water and formed the indentation that makes the thumb in the mitten shape of the Lower Peninsula. The north end of Saginaw Bay opens up into Lake Huron, and the south end is where the Saginaw River drains into the lake. This river is the principal river in a basin that drains water from the land at the base of Saginaw Bay into Lake Huron. The Tittabawassee River, to the west and north, drains water through what became the city of Midland and then into the Saginaw River. This area is the Saginaw Valley.

▪ ▪ ▪

By the mid-twentieth century, Michigan was a well-populated industrial hub best known for Detroit, Flint, and Saginaw—where Henry Ford, Walter Chrysler, and other automotive pioneers set the course for the history and economy of southeastern Michigan. But another completely different industry in the Lower Peninsula of Michigan can be found by driving from Detroit north about 115 miles on Interstate 75. While on Interstate 75, you will pass Flint and Saginaw, and then when you get to Bay City, turn west on US 10. After a quick nineteen-mile drive, you will reach Midland.

Midland is not like southeastern Michigan, and the man who made it different was Herbert Henry Dow, a chemist, not an automobile baron. Midland wasn't much of a town when he arrived in 1890, but it offered low-priced power in the form of lumbering waste as well as brine, from which chemicals could be extracted.

To understand Midland's people is to know that chemistry is its culture. The centrality of chemistry to the town's identity extends even to its students. Midland High School's athletic teams exemplify what Midland's people consider important. Their symbol is not an imposing, large athletic warrior like a Spartan or an aggressive, mean little beast like a Wolverine. Midland High's teams are called the Chemics, and their symbol is an atom.

Today the Midland exit off US 10 onto Business 10 leads you to the city center and Poseyville Road. A short drive south on Poseyville Road takes you to the Whiting Overlook Park. Stopping here and climbing the hill of the overlook gives you a view, but not a view of fields of flowers, the Tittabawassee River that flows through and occasionally floods Midland, or any other natural wonder. Instead, what you see is Midland's pride and joy: the gray and black skyline of the Dow Chemical complex.

If you worked for Dow Chemical here in the 1960s, when this story begins, there was a good probability that your father and perhaps even your grandfather had also worked for the company. In World War I the company's work with chemical synthesis became a vital military asset as supplies of chemicals from Germany disappeared.[8] And even before the United States entered World War II, Dow Chemical was supplying magnesium to the British, who were desperate for that material to build Spitfires and other military aircraft for the looming Battle of Britain. In 1942, a critical year for the production of aircraft, Dow Chemical produced 84.2 percent of all magnesium in the United States.[9]

By 1967, Dow Chemical was producing over eight hundred products that were sold world-wide. In a speech at that year's stockholders meeting, President Herbert D. (Ted) Doan boasted that Dow Chemical was the world's largest producer of chlorine, styrene, vinyl chloride, magnesium hydrate, bromine, and phenol.[10] Dow Chemical research produced a continuous supply of products that had provided jobs for generations of Midlanders. The *Midland Daily News* banner in 1967 read "Midland, Michigan—World's Greatest Chemical Center."

The research done in the town has always been a matter of community pride. The paper's front page on August 10, even carried a picture of the first nuclear reactor in Midland, which was located in Dow Chemical's radiochemistry research laboratory. The story below the picture reports that the reactor first reached "criticality"—a sustained chain reaction of fission of uranium 235—on July 6, with Dr. O. U. Anders at the controls. This was a small research reactor with a maximum power level of 100 thermal kilowatts (an amount equal to 134 horsepower) and was used primarily as a neutron source in Dow Chemical research.[11]

Midland's first nuclear reactor ran on the property of Dow Chemical, where 12,000 people worked and within a short distance of the city's schools and churches. The company had done everything it needed to do to conform to the Atomic Energy Commission requirements for construction and training. It then simply built the reactor, started it up, and featured it on the front page of the local paper. I could find no mention that anybody in Midland had any problem with having this small nuclear reactor in their midst. After all, Midland is a city that chemical innovation built.

But chemical production requires energy, and by the middle of twentieth century, the lumber waste that originally drew Dow Chemical to Midland was long gone. And because of the area's geological history, Michigan lacked large deposits of coal, oil, and natural gas. It was also too flat for large dams that could generate large amounts of electricity. In the 1960s Dow Chemical had coal delivered by rail to Midland, but Ohio coal was expensive and polluting. The company wanted an alternative.

In December 1967 the Consumers Power Company, the investor-owned utility that provided electricity to most of the Lower Peninsula outside of Detroit, announced its plans to build two huge nuclear reactors on the floodplain of the Tittabawassee River adjacent to the Dow Chemical plant and the City of Midland. These reactors would produce enough steam to run Dow Chemical's Midland operations, plus enough to produce electricity for over one million customers. And the reactors would do so at price equal to or less than if coal were used. With these two reactors in operation, it would not matter that Midland and the Saginaw Valley lacked large deposits of coal, oil, or natural gas, or that there were no sites available for large hydroelectric dams. The successful addition of nuclear power in Midland would be the answer to all of the area's energy needs.

In addition to providing power for Dow Chemical, building new power plants fit Consumers Power's view of what the state of Michigan needed.[12] As 1967 dawned everything seemed great in America: unemployment was below 4 percent, hourly wages were growing at 5 percent a year, and most Americans believed President Lyndon B. Johnson's story that America was fighting the war in Vietnam to keep the world safe from Communism.[13] In Michigan companies were building cars that people bought, and workers building the cars were being paid a good wage and receiving company-paid health insurance as well as a company-paid pension.

The great times were excellent news for Consumers Power. Its electrical sales had grown at an annual rate of 10.9 percent from 1957 to 1967.[14] In 1967, Consumers Power's industrial customers built 83 new manufacturing plants of major size, expanded 148 large existing plants, and added 174 smaller plants.[15] In their annual report that year, the company boasted that it supplied the energy used to manufacture a substantial portion of not only all cars and trucks but also the automotive parts and equipment made in the United States. Consumers Power believed that studies it conducted during 1967 demonstrated that their customers' demand would require the doubling of its generating capacity by 1975.[16] Midland appeared to be a fine place to locate additional electrical generation.

CHAPTER 2

Imagine This

A Whole Town Rallies in Support of a Nuclear Power Plant

The Consumers Power Company announced on December 14, 1967, that it would locate its nuclear plant across the Tittabawassee River from the Dow Chemical Company complex, where twelve thousand people worked, and adjacent to the city of Midland, where thirty-six thousand people lived.[1] The plan was to build two large nuclear reactors, with the first going online in 1974 and the second in 1975. The plant design called for the generation of 1.3 million kilowatts of electricity (enough to supply approximately one million homes). Additionally, it would deliver up to four million pounds of steam per hour via a bridge over the Tittabawassee River for industrial use at Dow Chemical.[2] The people of Midland welcomed the plan, and in this they shared many Americans' enthusiasm about the new source of power.

This nationwide enthusiasm for nuclear power emerged in 1957 when the first nuclear power plant dedicated solely for civilian use began producing electricity at Shippingport, Pennsylvania.[3] The Jersey Central Power & Light Company purchased a nuclear reactor for its Oyster Creek Plant in 1963. Jersey Central's analysis of the costs of this installation seemed to demonstrate that nuclear power could produce electricity at a cost equal to or less than coal.[4] Within five years of the Oyster Creek Plant order, American utilities placed orders for seventy-five new nuclear power plants.[5] One of the few skeptics of the alleged economics of nuclear power, Philip Sporn, president of American Electric Power, termed the utilities' rush to buy nuclear power plants in the late 1960s the "Great Bandwagon Market."[6] In hindsight, he would seem prescient.

Why was there so much interest in an alternative to coal? As early as 1958, Consumers Power believed that the future of electric utilities was nuclear power. Throughout the 1960s and the 1970s, executives were convinced that because the supply of coal was not adequate (even "dirty" coal that was full of sulfur was hard to get), oil and gas supplies were limited, and solar energy technology had not been developed, nuclear power was the answer for generating electricity.[7]

So the plan for two large nuclear reactors fit both Consumers Power's vision of the future and Dow Chemical's interest in finding a cheaper fuel source. The Midland company's January 1968 employee publication hailed the plans for these two large nuclear reactors as a great step forward: "This is without a doubt a proud achievement for Dow and the city it calls home, just as nuclear energy is a great accomplishment of mankind. But it is not an end in itself. It is a valuable hard-won instrument for progress. Its availability is more important to the accomplishments of tomorrow than to the achievement itself."[8]

But just a few years later in 1971 all of the enthusiastic plans for the Midland Plant were mired in a bureaucratic maze. Consumers Power had to stop construction in 1970 because the Atomic Energy Commission (AEC) had not yet issued a construction permit. In response, on the afternoon of October 12, 1971, the Midland Nuclear Power Committee (MNPC) staged a Speak Up Now rally in support of the plant. The sun was shining, the temperature was about 60 degrees, and there was a slight breeze. Dow Chemical encouraged its employees to take the afternoon off with pay to attend, and one hundred businesses closed, posting signs urging attendance at the rally at the Midland County Fairgrounds on the north edge of town.

The headline on page one of the *Midland Daily News* three days prior to the rally had read, "Challenge Given to All Residents to Attend N-Rally." The article quoted H. C. Allison, chairman of the Speak Up Now organization, "The rally is our single opportunity . . . to speak out in a united and forceful manner—loud and strong enough to be heard across this land and especially in Washington." His challenge and the support of the town's employers worked: the crowd that rallied to support the Midland Plant numbered between ten thousand and twenty thousand—about 40 percent of Midland's population.[9]

Financed by $32,000 ($200,000 in 2020 dollars) collected by the Midland

FIGURE 1. The Breadfruit Rock Band warms up the crowd for the Speak Up Now rally, October 12, 1971. (Courtesy of *Midland Daily News*)

Chamber of Commerce, the rally was a grand affair that featured local, state, and national celebrities.[10] The Breadfruit Rock Band, composed of local high school students Randy Vrchick, Charlie Todd, Jerry Meier, and Phil Coultrip, kicked off the event. The combined high school bands from H. H. Dow High School and Midland High School followed. The Berryhill Post 165 of the American Legion handled the flag raising. Dr. Arthur Turner, president of Northwood Institute, the local college, then introduced Robert Taylor, better known as Fat Bob the Singing Plumber from Detroit.[11] Fat Bob led the singing of the national anthem and then, as the master of ceremonies, introduced the speakers.[12]

Michigan's Republican senator Robert Griffin, along with thirteen other speakers (labor leaders, politicians—both Democrats and Republicans—and Chamber of Commerce members), voiced their support for the Midland Plant. Senator Griffin told the crowd that he would go back to Washington, D.C., with the story of Midland's demand that the construction permit be speeded up.

Then, in a a sign of emerging challenges to nuclear power, Mrs. Sidney N. Smock, a member of the MNPC, told the crowd about the recent federal court of appeals ruling on the Calvert Cliffs' Nuclear Plant in which the court had said for the first

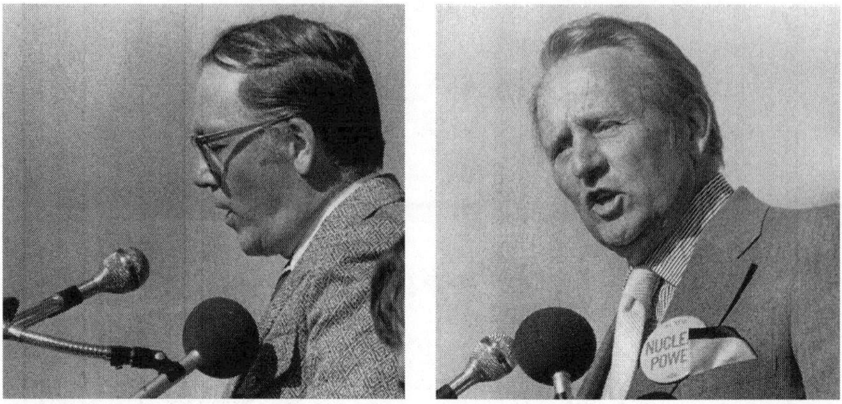

FIGURE 2. Senator Robert Griffin (*left*) and TV and radio personality Art Linkletter (*right*) telling the Speak Up Now rally crowd that they would take a message from Midland to Washington, D.C., that people in Midland wanted a nuclear power plant. (Courtesy of *Midland Daily News*)

time that environmental matters and cost-benefit analyses needed to be part of the AEC reviews for new plants. She said this court ruling was a setback for nuclear plants in the country, and she criticized the AEC for not holding full-time hearings, which further delayed proceedings.[13]

Next, TV and radio celebrity who had written the popular 1957 book *Kids Say the Darndest Things!*, Art Linkletter arrived on the fairgrounds by helicopter and told the excited crowd, "If the environmental people are so picky about the possibility of some little thing being disturbed and want to live by candle power and toast their bread on campfires, they are welcome to it but the rest of us want to go on living in the manner we are accustomed to." As a personal friend of President Richard Nixon, he promised to report to the nation's leader and those people in power in Washington about the Midland rally.[14]

And what could the thousands of people who attended the rally do to promote the Midland Plant? Mounted on a truck parked at the rally was a ten-by-twenty-foot sign for them to sign. The sign read "LICENSE FOR A NUCLEAR POWER PLANT GRANTED BY CITIZENS OF THE MIDLAND, MICHIGAN AREA." The plan was to deliver this sign to Washington with as many signatures as possible so that federal regulators and politicians would know that the town wanted its nuclear reactors.[15]

The last item on the rally program was to parade this sign as the crowd sang "We Need Nuclear Power Now" to the tune of the "Battle Hymn of the Republic." The lyrics they sang on that beautiful October day in 1971 were

FIGURE 3. The "License for a Nuclear Power Plant" available to sign at the Speak Up Now rally to show support for the Midland Plant. After the rally, this sign was taken to Washington, D.C., with fifteen thousand signatures. (Courtesy of *Midland Daily News*)

We have gathered here together to assure a future bright;
To get power for our jobs each day and light our lamps at night.
For ourselves and for our children what we're asking for is right,
We need nuclear power now.

Glory, Glory, Hallelujah
Glory, Glory, Hallelujah
Glory, Glory, Hallelujah
We need nuclear power now.

Cleaner air and water for the mid-state is our stand,
For the welfare of our people and the future of our land. Let us tell the folks in Washington a license we demand,
We need nuclear power now.
(Repeat chorus)[16]

Why did all of these Midlanders turn out for the Speak Up Now rally? Dick

Miller, president of the Midland Chamber of Commerce, said on the television show *Sixty Minutes*, "And the reason we had that extravaganza that we did was nothing more than to tell the world we wanted that plant."[17] It seemed like a simple request that afternoon in 1971. It was time to embrace the new miracle technology of nuclear power, get rid of the dirty coal plants that Dow Chemical was running in Midland, and save jobs in the town by producing steam and electricity at a cost equal to or less than coal-fired plants. The huge turnout for the rally seemed to validate the desire for nuclear power.

But not everyone in Midland wanted the construction permit hearing to rush to a positive conclusion for Consumers Power. One of the most interesting statements following the whole Speak Up Now rally affair came from a high school student, one of the one thousand students among the crowd. Charles Overberger, then at Dow High School, told the *Midland Daily News*, "Personally I think the people of Midland are more concerned with money than with pollution. I don't think the rally is going to pressure the AEC to make an early decision anyway. I also think it is sensible to go through all the formality of the 'red tape' in obtaining a license for construction of the plant. It's a necessity for the safety of the town. It's not that I'm against nuclear power: it's just that I don't think it should be pushed through this way."[18]

Overberger told me he had been part of the environmental movement then and even earned an award for his participation. He remembers that nobody talked about nuclear pollution in Midland, although he thought everybody knew about the hazard. He thought waste from nuclear energy was a big problem.[19]

It turned out that fifteen year-old Overberger predicted what would unfold over the next several years. In spite of the huge turnout at Speak Up Now and the speakers' promises to apply political pressure, the effort to speed up the granting of the construction permit failed. The Midland Plant did not get its permit until December 15, 1972, two years after the start of public hearings and five years and one day from when Consumers Power had announced the plant.

CHAPTER 3

Regulators

Safety Rules and the Spanish Inquisition

The Washington bureaucrats presiding over the red tape referenced by Charles Overberger were the regulators who worked for the Atomic Energy Commission (AEC). The AEC had been created by Congress as part of the Atomic Energy Act of 1954 when there was a broad political consensus for the development of nuclear energy for civilian purposes. This act assumed that safe reactors could be designed to heat water for steam to generate electricity, and it allowed the AEC to develop and enforce regulations that would at times rein in the industry in order to ensure the safe, peaceful use of nuclear energy and protect the public health and safety from radiation dangers. However, the act also gave the AEC the conflicting role of acting as a cheerleader promoting the use of nuclear energy.[1]

Midlanders, in their attempts to use whatever political clout they could muster to expedite the approval of the construction permit for the Midland Plant, had to try to deal with the AEC's ongoing efforts to develop and administer regulations for nuclear power. Matters were made worse by a problem of Consumers Power's own making. Although the company had started planning its nuclear installation in 1966 and announced it in December 1967, it did not file its application for a construction permit until January 1969.[2] By the time it was finally filed, the AEC was experiencing a licensing and inspection caseload increase of about 600 percent between 1965 and 1970 because of the rapid increase in the number and size of reactors ordered by utilities.[3]

▪ ▪ ▪

The AEC staff struggled in the late 1960s and early 1970s with gaining the public's confidence that they understood the risks of nuclear power generation.[4] Americans' introduction to nuclear power had been its use for mass killing in World War II. On August 6, 1945, in a bid to end the war, the United States had dropped the world's first atomic bomb on the city of Hiroshima, Japan. With an equivalent force of 12,500 tons of TNT, this bomb, called Little Boy, killed at least 130,000 people, and another 70,000 would die from bomb-related injuries by 1950. Three days later, the United States dropped a second bomb, called Fat Man, with a force equivalent to 22,000 tons of TNT, on Nagasaki, Japan. An estimated 140,000 would die from this second bombing.[5]

Wilfred Burchett, the first journalist to visit Hiroshima after the atom bomb was dropped, arrived alone by train from Tokyo on September 2, 1945. His Morse code dispatch, "The Atomic Plague," printed in the London *Daily Express* three days later, was the first public report to mention the effects of radiation and nuclear fallout.[6] He wrote, "In Hiroshima, 30 days after the first atomic bomb destroyed the city and shook the world, people are still dying, mysteriously and horribly—people who were uninjured in the cataclysm—from an unknown something which I can only describe as the atomic plague."[7]

Beyond the immediate deaths from radiation, the long-term effects on the people of Japan were extensively studied in the years following World War II. In September 1967 the *Detroit Free Press* reported under the headline "The Stunted Children of Hiroshima" that radiation from the atom bomb blasts at Hiroshima and Nagasaki had stunted the physical and mental growth of many unborn babies whose mothers had been near the blast sites. They cited the findings of the Atomic Bomb Casualty Commission, a U.S.–Japanese cooperative research agency that had studied children exposed to the radiation in utero up to age seventeen. Dr. James W. Wood, the author of this report, also published the results of his studies in the American Academy of Pediatrics publication *Pediatrics*.[8]

Aware of this public perception of atomic power's danger, the AEC tried to make certain that its staff understood the radiation risk from nuclear power plants and provided regulation to mitigate it. One utility executive called the resulting licensing process "a modern Spanish Inquisition carried out by AEC engineers, scientists, and consultants [who] have no serious economic discipline."[9]

▪ ▪ ▪

Although the AEC did attempt to streamline its licensing procedures, the commission found it impossible to reduce its review time. Part of the reason was that in the late 1960s and early 1970s, the AEC staff began to have doubts about some of the safety features of nuclear reactors. The issue of greatest concern by the late 1960s was a meltdown of the core of a nuclear reactor that could be caused by a loss of cooling water. Because of this specific issue, the AEC stopped the Midland Plant's construction permit hearing for ten days so that its employees could attend the separate hearings on emergency core cooling systems. In addition, the AEC instructed Consumers Power to redesign the emergency core cooling system for the Midland Plant. Then it took the AEC at least six weeks to evaluate this new design.[10]

Starting in the late 1960s, the AEC also had to try to overcome the general disillusionment with the government, established institutions, and science that prevailed largely as a result of the Vietnam War. One college student summarized the situation after listening to a Brookhaven National Laboratory expert on radiation: "Dr. Bond sounds good, but we can't believe him. He works for the Government."[11]

▪ ▪ ▪

The construction permit hearings were also time-consuming because of all the people allowed to speak in them, a group officially known as "intervenors." A decade before the Midland Plant's hearings, construction permit hearings were not public, but these closed-door hearings were a thing of the past by the early 1970s. And the reason why individuals or groups opposing the construction of a nuclear plant got the opportunity to participate in construction permit hearings is a Michigan story.

In Michigan, work had already started to use atomic energy for industrial processes prior to the 1954 Atomic Energy Act when a study initiated in April 1951 by the Detroit Edison Company and Dow Chemical had led to the formation of the Power Reactor Development Company (PRDC).[12] The AEC approved a construction permit in 1956 for PRDC to build the experimental Enrico Fermi Atomic Power Plant, Unit 1 (Fermi 1) south of Detroit. The placement of Fermi 1 close to urban areas solved the sticking point of high transmission costs from remote sites.

Walter Reuther, the longtime leader of the United Auto Workers union (UAW), was one opponent. He had gained his reputation of looking out for workers with the

FIGURE 4. Walter P. Reuther, president of the United Auto Workers union (UAW) from 1946 until his death in 1970, at the 1966 UAW Constitutional Convention. In the 1950s Reuther opposed nuclear reactors being located near where people lived, and on August 31, 1956, labor unions filed a suit to block the issuance of a construction permit for the Fermi 1 nuclear reactor. (Courtesy Walter P. Reuther Library, Archives of Labor and Urban Affairs, Wayne State University)

1950 labor contract he achieved between the UAW and auto makers called the "Treaty of Detroit." This agreement affected the lives of many laborers and influenced labor contracts and law making in the following decades. It made pensions and healthcare a permanent part of such agreements, moved American autoworkers into the middle class, and made home ownership common among laborers.[13]

Not just concerned strictly with labor contracts with employers, Reuther also had a wide ranging interest in all sorts of issues beyond the job site of his workers. He thought nuclear reactors were dangerous and should not be built in urban areas. In opposing the Fermi plant, he said the location of the Fermi reactor thirty miles south of Detroit and thirty miles north of Toledo in an area where five hundred thousand UAW members lived exposed all of the union members, as well as the general population, to too much risk.[14] Following words with actions, the UAW, the International Union of Electrical Workers, and the United Papermakers and Paperworkers challenged the AEC's issuance of the construction permit it had approved for Fermi 1. The unions pursued this case, *Power Reactor Development Co. v. Electricians*, all the way to the Supreme Court.[15] They lost their case on June 12, 1961, when the court ruled in favor of the AEC.

But the unique fast breeder design of the Fermi 1 reactor, which used liquid sodium as a coolant, suffered from multiple problems and was the last one of that type completed in the United States. The risks were dramatically illustrated in 1966 when a problem with the cooling system caused the fuel rods to melt down, creating a radiation leak. This accident was described in 1979 as the most serious nuclear accident that had happened in Michigan.[16]

Some have speculated that Reuther may have also been motivated by his desire for public entities—rather than private investor-owned utilities—to build nuclear plants.[17] But, regardless of what his motivations were—the risks of the fast breeder design, the problems with the site, or the fact that Fermi 1 was not being built by the government—Reuther and the unions made a difference in the requirements to build nuclear reactors. The unions' protests over the granting of the construction permit for the Fermi 1 were the driving force behind a change in the Atomic Energy Act of 1954 whereby AEC reports had to be made public and public hearings had to be held for construction permit and operating license applications.

Gerard H. Clarfield and William M. Wiecek summed it up as follows:

> the UAW established the principle that nonparties in regulatory hearings have the right to participate as "intervenors." This effectively opened AEC hearings to public participation. Previously, such hearings had been held behind closed doors, with only the AEC's decision, not the testimony, printed in the *Federal Register*. Now AEC hearings became public forums, open to the unions, environmentalists, and activists of all sorts, much to the AEC's chagrin. The new role of intervenors also turned AEC hearings into legalistic struggles, where administrative hearings were replayed in the courts after dissatisfied parties or intervenors sought to have a judicial authority overturn some regulatory rule.[18]

■ ■ ■

In sum, three factors contributed to the time it took to get a construction permit for the Midland Plant: (1) the questioning by people who opposed the Midland Plant in the construction licensing hearings made possible by Walter Reuther's earlier protests, (2) understaffing at the AEC, and (3) the AEC staff's concern that they did not thoroughly understand how the cores of nuclear reactors should be cooled. But as will be detailed later, for the entire time the Midland Plant was under construction, the AEC (after 1974 the Nuclear Regulatory Commission), continued to struggle to fulfill its statutory mandate to make sure the Midland Plant would be safe.

CHAPTER 4

To Build a Nuclear Power Plant

Admiral Rickover's Way

Whereas Walter Reuther opposed the operation of nuclear reactors in close proximity to where people lived and worked, others pointed to the use of nuclear power for ships and submarines, where sailors would sleep next to nuclear reactors. If you could make nuclear reactors work in ships and submarines, then why couldn't you use nuclear reactors to generate electricity for civilian use? In 1967 Consumers Power Company, with its announcement of the Midland Plant, said of course you could—and should—use nuclear power to produce steam to make electric power as well as provide steam for the Dow Chemical Company.

Consumers Power's enthusiasm for nuclear power stemmed in part from developments in reactor design and demonstrations that nuclear power could indeed safely produce electricity. The company selected not the sodium-cooled reactor design used for Fermi 1, but instead a model that used ordinary water as a neutron moderator, coolant, and heat transfer material.[1] Admiral Hyman Rickover led the effort to use the water design in ships and submarines as well as for civilian use. His work led to both the 1954 launch of the USS *Nautilus* (*SSN-571*) nuclear-powered submarine and the 1957 start-up of the Shippingport Atomic Power Station in Pennsylvania.[2] Perhaps even more important, Rickover showed

FIGURE 5. In August 1958 Admiral Hyman Rickover received the Congressional Medal of Honor in recognition for successfully directing the development and construction of the world's first nuclear-powered ships and the first large-scale nuclear power reactor devoted exclusively to the production of electricity. (Courtesy of the Naval History and Heritage Command now in the collection of the National Archives, official U.S. Navy photograph of Rear Admiral Hyman G. Rickover, USN. Portrait photograph taken circa mid-1955.)

how to manage the complex task of building a nuclear power plant while assuring that it can be safely operated.

The world's first human-made nuclear reactor, called Chicago Pile-1, had been constructed under the direction of Enrico Fermi beneath the west wing of the old Stagg Field football stadium at the University of Chicago in 1942. Fermi, a genius of both theory and experimentation, was successful in creating the first controlled nuclear chain reaction of the fissioning of atoms.[3]

On the morning of December 2, 1957, fifteen years to the day after Fermi had achieved that first nuclear chain reaction, the Shippingport reactor, the first full-scale nuclear power station designed solely for commercial power purposes, went into operation. This plant had the ability to generate 60 megawatts (MW) of electricity. An often used crude measure estimates that a 100 MW capacity can supply one hundred thousand homes, though the actual number depends on how much of the time the generating plant is in operation and several other factors.

Then only ten years after the 60 MW Shippingport reactor started up, Consumers Power announced that it would construct its 1,600 MW nuclear facility in Midland.[4] Not only would their facility produce approximately twenty-six times more power than Shippingport, it would have the additional complexity of a system that would also supply steam to Dow Chemical.

Denominated in Michigan terms, advancements in nuclear power were super-charged at a rate that far exceeded the rate of advancement of power in the auto industry. In about 1870, Austrian inventor Siegfried Marcus was the first person to attach an internal combustion engine to four wheels and make a "car."[5] It was not until 1908, that the first affordable car rolled off the assembly line at the Ford Motor Company in Detroit, the Model T Ford, with a 20 horsepower engine. By the 1960s Detroit automakers produced "muscle cars" with 400 horsepower engines. In contrast, only thirty-three years would elapse between Fermi's first demonstration of controlled nuclear fission in 1942 with only a few watts of thermal power and the projected 1975 completion date of Consumers Power's Midland Plant, which would be able to produce 4,936 MW of thermal power.[6] Midland Plant's thermal power would result in the production of approximately 1,300 MW of electrical output, plus the steam for Dow Chemical's use (the equivalent of approximately 300 MW of electric power). The difference between the thermal power generated and the equivalent electrical output of the plant is accounted for by the fact that a good deal of the heat produced by a nuclear reactor, as in conventional power plants, is not converted into useful electric energy.

How was it possible for the development of nuclear power to advance so quickly to provide steam for useful purposes? This can be attributed to Admiral Hyman Rickover, who became the "father of the nuclear navy." The mere fact that Rickover had a sixty-three–year career in U.S. Navy—let alone that he was a pioneer in nuclear power—is remarkable. When newscaster Diane Sawyer interviewed him late in life he said, "I have the charisma of a chipmunk. I never have thought I was smart. I thought other people I dealt with were dumb, including you."[7] How was it, then, that this tactless man rose to prominence in the United States Navy?

Besides possessing an acidic personality, Rickover was not an impressive physical specimen. From my one brief encounter with him in 1962, I remember him as a foul-mouthed little man (his height was reported as five-foot-six and his weight as 130 pounds).[8] At this meeting he was dressed in a civilian coat and tie. His white shirt appeared to have a neck size about two inches too big, and he appeared to me to be a weaseled-up old man. (At age twenty-two, I had little knowledge of or appreciation for Rickover's accomplishments.)

Rickover had no family background in the U.S. Navy. He was born in Poland, then part of Russia, to Jewish parents. His father immigrated to the United States and subsequently saved enough money to send for his wife, daughter, and Hyman. Rickover's appointment to the Naval Academy came as the result of an uncle's political connections.[9]

Rickover went to work at age fourteen as a Western Union messenger, working from 3 p.m. to midnight, and was able to get a high school education only by scheduling his classes around his work hours.[10] His academic preparation for the Naval Academy consisted of studying on his own for the entrance exam after determining that the preparatory course he had paid for was worthless. Thus, he arrived at the Naval Academy poorly prepared, and his first year was complicated by the fact that he was quarantined for several weeks as a possible diphtheria carrier.[11]

Rickover was usually not well received by his peers. In order to make up for his lack of academic preparation, he rigged up a place to study in the shower after taps. This practice of studying after hours did not sit well with his classmates as they considered it unsportsmanlike.[12] Also, his reception was not enhanced by the fact that he was Jewish at a time when anti-Semitism was rampant.[13] After he graduated from the academy in 1922, getting to know his fellow officers was not a high priority for him. When the submarine (*S-48*) he was serving on spent time in harbor in the Panama Canal Zone, he elected to spend time with his wife rather than joining his shipmates in card games or in drinking sessions ashore.[14]

The standard career path of a navy officer includes having a successful tour of duty as the commanding officer of a ship. As an engineer officer, Rickover received good fitness reports (personnel evaluations), but these assignments did not lead quickly to the coveted command at sea. Although he was second in command of the *S-48*, he was passed over to become its captain. The submarine high command did not consider him the easygoing, swashbuckling command figure that other officers and men would prefer to serve under.[15] Rickover was finally dispatched in July 1937 to be the commanding officer of a rusty old tub, an aging minesweeper, the USS *Finch*. However, before reporting, he had applied for "engineering duty only" status, apparently convinced that his assignment to the *Finch* indicated that pursuing the normal career path was not going well. This change of status came through in the fall of 1937, and he therefore only spent three months as the commander of a vessel.[16]

What Rickover did possess was a deep appreciation for education, an obsession with the proper construction of equipment, and the ability to solve complex engineering and organizational problems. Because he had arrived at the Naval Academy ill-prepared, he had to find a way to educate himself and in the process developed a lifelong interest in learning. He later criticized the education he received at the Naval Academy for placing too much emphasis on memorization and not enough time on how things worked. However, the Navy did send him to Naval Postgraduate

School and to Columbia University, where he earned a master's degree in electrical engineering and developed a deep appreciation for the way things work.

Although Rickover's career took him out of the normal-command-at-sea career path, his assignments prepared him well for his later role in nuclear power. During World War II, he served as the head of the Electrical Section of the Bureau of Ships, which was responsible for building and maintaining the Navy fleet. When he was given full charge of the section in March 1941, he built a team that had the technical expertise to evaluate and demand the equipment that the Navy needed. Instead of the previous practice of allowing manufacturers to set technical specifications for Navy equipment, Rickover saw his service as the customer as having the right and responsibility to set specifications and see that the manufacturer met them. He also demanded that contractors actually delivered equipment on time. He became famous for his "shock resistance" test—he called it the "radiator test"—which consisted of throwing the equipment against the radiator to see if it survived.[17]

In July 1946, having received the Legion of Merit for his efforts leading the Electrical Section of the Bureau of Ships, he was assigned along with four other naval officers to Oak Ridge, Tennessee, to take a yearlong course in nuclear science and technology.[18] Losing no time, he called a meeting of all of the Naval officers and told them that, as the senior officer, he would be writing the fitness reports of the other four and that he would define their course of study. Thus he was able to control the Navy's first efforts to deploy nuclear power.[19]

By 1948 and 1949 he had successfully maneuvered dual assignments as director of the Navy's Nuclear Power Branch (designated Code 390) and a position in the Atomic Energy Commission's (AEC) reactor development organization.[20] This twin role enabled him to lead the effort to develop the Navy's first operational nuclear-powered submarine, the USS *Nautilus* (*SSN-571*), which was launched and commissioned in 1954. It also enabled him to oversee the development of the Shippingport Atomic Power Station, the first commercial light-water pressurized (PWR) power plant in the United States. One of his important decisions in the design of the Shippingport reactor was that the fuel elements would be made of uranium oxide clad in zirconium-alloy tubes.[21] This system is used to this day in nearly all the world's reactors, and only since 2019 have a number of other designs been proposed.[22]

Rickover's involvement in the Shippingport project placed the United States on a path to use light water reactor systems for civilian power generation.[23] He and the Naval Reactors Branch directed all aspects of its design. Although Westinghouse, as the prime contractor, was theoretically responsible for coordinating all construction and procurement, in reality Rickover installed his own people to uncover problems and propose solutions. This constant threat of his intervention was always a stimulus at Shippingport.[24]

When he was assigned the project in the summer of 1953, Rickover set the fall of 1957 as the date the plant would be completed. The reactor went critical (reached a state in which a nuclear-fission chain reaction became self-sustaining) on schedule on the morning of December 2, 1957, fifteen years to the day after Enrico Fermi had achieved that first nuclear chain reaction in Chicago. At 11:10 a.m. on December 23, the pressurized-water reactor reached its full net power rating of 60 MW of electricity.[25] This reactor at Shippingport was the start of the journey that led to Consumers Power's proposal to build nuclear reactors in Midland.

In August 1958, Rickover received a Congressional Gold Medal in recognition for successfully directing the development and construction of the world's first nuclear-powered ships and the first large-scale nuclear power reactor devoted exclusively to the production of electricity.[26] In testimony provided to the Joint Committee on Atomic Energy in June 1961, he reported that he had been responsible for the construction of forty-four nuclear power reactors. When asked how many people (both civilian and military) had been employed in those projects, he estimated one hundred thousand. No accidents had happened at any of the reactors, and no personnel had absorbed any unusual dose of radiation.[27]

Why was Admiral Rickover so successful, first in his job at the Bureau of Ships during World War II and later with the building of nuclear reactors? In a speech at the U.S. Naval Postgraduate School in March 1954, he said,

> In 1945, after the defeat of Germany we sent technical missions to Europe to learn what we could of the enemy war effort . . . Col. Leslie Simon of the U.S. Army reported that the German Air Force had been far superior to the German Army and Navy in science and in technology because the German Air Force had available at their headquarters organization people who were equally competent as those in industry with whom they had to deal. The lesson learned from this is that a military agency which aspires to do a large development job, and rapidly, must have at its headquarters scientists and engineers just as competent as those who are doing the work in the field. Otherwise, the headquarters is at the mercy of the field and becomes essentially an agency for merely supplying the necessary funds and rubber stamping the technical decisions.[28]

It would have been well for the Consumers Power executives and board members at the time this story takes place to have been familiar with Rickover's speech.

Milton Shaw, who worked first in Rickover's naval propulsion program and later served as the AEC's director of Reactor Development, summed up the Rickover approach: the key to the technical success of the submarine propulsion plants was *governmental engineers* with absolute control of industrial contractors. The Naval Reactors Program was a "government program" in the sense that technically expert government administrators exercised daily control over the activities of their contractors.[29]

The most frequently noted points about the significance of Rickover's Shippingport Plant are that it demonstrated that nuclear power could be used for civilian purposes; every aspect of its design was declassified, which meant that other companies could learn the rudiments of nuclear technology; it proved the success of the pressurized water design for nuclear power; and its initial operation was flawless, including its response to fluctuations in electrical demand. Lost in many accounts is the fact that it was built on schedule with Admiral Rickover's retaining almost absolute control of the construction management.

According to the Consumers Power engineers that I interviewed, the company did not have the staff to maintain Rickover's style of absolute control over the myriad industrial contractors necessary to build the Midland Plant. Will Davis's history of the nuclear construction cost overruns and delays in the 1970s describes the general problem: the construction of early nuclear plants required the prospective owner operator (an electric utility company) to contract for a detailed design of the nuclear plant with an architect-engineer firm, for the construction of the nuclear plant with one or more construction contractors, for a nuclear steam supply system with a reactor vendor, for a turbine generator also with a vendor, and for the rest of the various other pieces of equipment with further contracts. Management of all these various contractors, and all of their innumerable subcontractors (as well as thousands of workers in many locations) could get complicated and messy. The fact was that only a few utilities, such as the Tennessee Valley Authority, Duke Energy, and Detroit Edison, had the ability to perform major portions of this work entirely in-house.[30]

In spite of the challenges of construction management Consumers Power faced, the company executives thought they had the ability to solve Dow Chemical's need for power and also to harness nuclear power for all of their new power generating plants.

CHAPTER 5

A Utility and a Chemical Company

Can This Marriage Endure?

Every three days in the 1960s, ninety-five train cars, each car loaded with 110 tons of coal, left southern Ohio bound for Midland.[1] The Dow Chemical Company burned 1.3 million tons of coal each year in Midland to produce the necessary steam and electricity for its chemical operations, laboratories, and offices.[2] By the 1960s the company had grown and changed since 1890, when founder Herbert Henry Dow had come to Midland with $100 cash, a bank draft for $275, and an idea about how to recover bromine by electrolysis from brine.[3] Dorothy Langdon Yates described what his brother-in-law, Thomas Griswold Jr., a chemical engineer, found when he joined Dow in Midland seven years later:

> The scene in Midland as Griswold reported it was of a "dead lumber town." On his arrival the river had overflowed the flats and Main Street was an expanse of either mud or dust. The wind blew the horse manure onto the fruits and vegetables displayed for sale outside stores. No decent woman dared linger on the east side of Main Street, the site of 14 saloons. Dry goods stores were on the west, or river side.
>
> A woman wishing to buy a spool of thread had to wade through the mud or dust of Main Street. The pine sidewalks, pockmarked by lumbermen's boots, were crumbling. Occasionally someone would fall through and break a leg.

> The air was alive with bugs. The noise of crickets, katydids, and frogs made it hard to sleep. Outside the city were swamps and millions of pine stumps overgrown with vines, sticking up as a graveyard of fire-blackened stubs.[4]

It should be noted also that the Tittabawassee River was not like Hemingway's sparking clear rivers of northern Michigan, rivers that make you imagine beautiful brown, rainbow, and brook trout shimmering just below the surface. Most of the year the Tittabawassee is a slow-moving, muddy river that brings to mind carp, suckers, and bullheads.

The wealth from the lumbering boom in the Saginaw Valley in the 1880s and 1890s resided in neighboring Saginaw and Bay City. In 1890, sawmills on the Saginaw River, which flows from Saginaw through Bay City, cut 815 million board feet of lumber. In 1887, thirty-two sawmills that cut over four hundred million board feet of lumber were operating on the Bay City section of the Saginaw River.[5] In contrast, on the Tittabawassee only one sawmill was operating in 1890, and it manufactured barrels for locally produced salt.[6]

In the late 1800s, the lumber barons built their Victorian mansions on Center Avenue in Bay City and in downtown Saginaw. In contrast, Midland houses from the period, with only a few exceptions, were modest wood-frame structures. In the 1880s and 1890s the Presbyterians and the Episcopalians in Saginaw and Bay City built imposing gothic stone and brick churches; in Midland the Presbyterians worshipped in a modest brick structure and Episcopalians in a building that had been converted from John Sias's barn.

So it was not that Midland offered great natural beauty or a thriving urban atmosphere for young professionals that brought Dow to Midland in 1890. What Midland had to offer was plentiful fuel for a price he could afford and brine rich in bromine that could be pumped from wells. He had developed a chemical process to recover bromine from brines, and his brine recovery process required electricity. Midland in 1890 had ample fuel available from lumbering waste, and shortly thereafter coal became available from mines just to the east of Midland around Bay City and Saginaw. But this fuel would not last forever.

In 1890, Dow, who had the chemical genius, and John H. Osborn, who had $3,000, formed the Midland Chemical Company.[7] This company produced bromine using Dow's process.[8] Dow had a falling out with the other parties involved in this company in 1893 and left it to work on a new invention. By 1896, he had perfected a way to produce chlorine from brine by means of an electrolytic cell, which allowed for the production of chloride of lime, or bleaching powder. He formed a new company to produce it, the Dow Chemical Company.[9]

The availability of fuel in Midland was an important part of Dow's 1897 economic justification to raise the capital for Dow Chemical. He described his bleach plant as needing 60 tons of coal per day. He noted that coal of unusually good quality could be obtained for seventy cents per ton from three mines that had opened within the last year in the Saginaw Valley, and mentioned that other mines were being started.[10] It appears that one of these new mines was the Monitor Mine, which was approximately ten miles east of Midland and started operations in 1896. This mine is reported to have produced high quality bituminous coal at a rate of three hundred tons a day.[11]

Dow was correct that other coal mines would open in Michigan, at least in the short term. The greatest tonnage of coal produced in Michigan was in 1907 when a total of two million tons was obtained from thirty-seven mines. However, because of the state's geology, these beds were usually less than three feet thick and less than 150 acres in extent.[12] By 1923 coal production had declined to less than a million tons a year, and in 1949 it was approximately twelve thousand tons from only one mine. The last coal mine in Michigan closed in 1950.[13] Fortunately for the company, after the lumber waste had disappeared and as local coal beds played out, Midland County had an oil boom in the 1930s and 1940s. During those decades the company burned fuel oil.[14]

But by 1967 Dow Chemical had a problem. The lumber waste in the Midland area that had provided the fuel to make electricity had long since all been burned; Michigan's coal mines were all shut down; Midland County's oil boom of the 1930s and 1940s was history with the production from these fields reduced to small amounts; and the state's other oil and natural gas deposits did not offer an inexpensive source of fuel. Hence in the 1960s Dow Chemical's fuel to make steam and electricity for its Midland facilities was the coal delivered by rail from Southern Ohio. Macauley Whiting, Dow Chemical board member and director of basic resources, wrote in 1968 that the company had several hundred million dollars invested in its Midland chemical complex, but that fossil fuel costs in Midland were very high relative to other locations in the United States.[15] These facts put the company in a difficult position.

Back in the 1930s, Willard Dow and Dutch Beutel (the top two executives in Dow Chemical at the time) had recognized the possibilities of low-cost energy for chemical plants in Texas when they saw natural gas being "flared off" just to get rid of it in the production of oil. In 1940, Dow Chemical built a plant in Freeport, Texas, to take advantage of it, and in the 1960s, the state continued to be an attractive low-cost energy alternative to operations in Midland.[16] But was it worth abandoning the company's investment and its twelve thousand employees in Midland?

In 1978, 60 percent of Dow Chemical's production was commodity chemicals (basic substances that were freely interchangeable with the same chemical produced by other manufacturers) that were sold to other chemical companies to make finished products. Most of the remaining 40 percent consisted of things like weed killer, brake fluid, and antibiotic ointments. In 1978 consumer products made up only 3 percent of Dow Chemical sales, the best-known being Saran Wrap.[17] Profitably selling commodity chemicals depends on having the lowest cost of manufacturing, which in turn requires the lowest cost of power. Cheap fuel was also necessary for research operations in Midland, to heat labs and operate its pilot plants for developing new products.[18] Zoltan Merszei, president and chief executive officer of Dow Chemical, put it this way in 1978, "Energy is not just power—it is our raw material, our competitive edge and our profitability."[19]

Not only was the price of producing electricity and steam from Dow Chemical's coal plants in Midland too high in the 1960s, their aging facilities created an incredible amount of pollution. As a result, the town was blanketed with coal ash, soot, and smog. Marge Mesler, who lived there at the time, remembers how almost every morning everything was covered with coal ash. This was not exactly conducive to the "sparkle city" image that Dow Chemical needed to facilitate its recruitment of the best scientists and engineers, but local people understood what it meant. Mesler said they called it their bread and butter.[20]

Thus, the benefits of the Midland Nuclear Plant for Dow Chemical appeared to be twofold: steam at a cost that would be competitive with the steam cost in Texas and the elimination of the coal mess in Midland. Building the nuclear plant would give Consumers Power an opportunity to continue its ambitious plans to add more generating capacity at a cost it said would be four-tenths of one cent per kilowatt hour, which would be less than power from a coal-fired plant located in Midland cost.[21]

But the Consumers Power announcement of the plan in December 1967 left it unclear who had come up with the idea. The press releases said only that Consumers Power and Dow Chemical had previously engaged the Bechtel Corporation to do a year-long feasibility study to determine if a dual purpose plant—electricity for Consumers Power's customers and steam for Dow Chemical—would be economically desirable for both parties. The 1967 announcement indicates that Bechtel's study had convinced Consumers Power that the price of electricity produced by the Midland Plant would be acceptable as well as had given Dow Chemical confidence that the price it would pay for steam would be competitive. The plant would also provide the added benefit of ending Dow Chemical's coal-burning pollution in Midland.

Who was it that that proposed this idea of a dual purpose plant in Midland? A 1977 article said it was Dow Chemical's idea.[22] Former Dow Chemical chemist, Dr. Robert Asperger remembered a meeting with Herbert D. (Ted) Doan in the mid-1960s in which Doan had expressed a keen interest in a plan in Europe to use a nuclear plant for power at a chemical complex.[23] The BASF (Badische) Chemical Company had announced in 1966 that it was examining the possibility of using nuclear energy in a power station at Ludwigshafen, Germany, that would supply not only electricity but steam as well.[24]

Doan would certainly have had the clout and the connections to kick off the Bechtel study for the Midland nuclear plant project. The grandson of the company's founder, he had been manager of the chemical department before becoming an executive vice president in 1960 and then following his father as chief executive officer in 1962. In *Growth Company*, Ned Brandt reports that Doan's father, Leland I. (Lee) Doan, had received a personal letter from S. D. (Steve) Bechtel, chairman and president of the Bechtel Corporation, congratulating him when his son was promoted to executive vice president in 1960.[25]

One thing is certain: whoever proposed the idea to site the nuclear plant in Midland, Dow Chemical's executives in 1967 had unmitigated praise for the plan. The *Midland Daily News* reported on December 14, 1967, that Doan said, "Dow is delighted to have entered into the contract." Doan went on to say that Dow Chemical was getting out of the power business, and he added: "You find sometimes in a new technology an area where you can't do as well yourself." The conclusion of the article reported that the company's president saw nuclear power as suited to the future needs of Dow Chemical, and with it, the start of a dream.[26]

Carl A. Gerstacker, Dow Chemical's board chairman, said, "Our agreement with Consumers Power represents a significant departure from the historic Dow position of producing our own power at major production complexes. This should serve to emphasize Dow's interest in working with utilities and buying energy when it can be shown this is more economical than operating our own power-generating facilities."[27] Harold Bosscher, general manager of the Dow Midland Division, said

FIGURE 6. December 14, 1967, James H. (Jim) Campbell (*left*), president of Consumers Power Company, and Herbert D. (Ted) Doan (*right*), president of the Dow Chemical Company, sign an agreement whereby Dow Chemical would purchase steam and electricity from Consumers Power's nuclear facility. (Courtesy of the *Midland Daily News*)

that the plans for the nuclear energy plant were "very important and significant for the long-term growth of the Dow Chemical Company's Midland plant and the expansion of employment opportunities."[28] He told the January 1968 Dow Chemical employee newsletter that "development of resources and raw materials will be aided by the availability of steam at a competitive cost. The new plant will supply up to four million pounds of steam per hour. This will be particularly advantageous in the manufacture of inorganic chemicals which require large amounts of steam."[29]

Not only did the design of the Midland Plant have the capacity to serve all of the steam requirements of the Dow Chemical complex in 1967, it would also have the capacity to provide the company with 25 percent more steam than it was presently using in the coldest part of the winter. While heating buildings was the most common application of the steam, processing chemicals accounted for the greatest volume. For example, in 1967 caustic soda and Dowflake calcium chloride both required large amounts of steam to manufacture. Indeed, the steam needed to produce calcium chloride alone exceeded the amount needed to heat the entire Dow Chemical complex of eight hundred buildings.[30]

The contract Dow Chemical signed in 1967 with Consumers Power for power from the Midland Plant was the largest contract the chemical company had ever signed.[31] James H. (Jim) Campbell, Consumers Power president, said, "We expect a long and happy marriage between our two companies. This is the fruition of a joint effort to bring technology to an entirely new level. The plant is the result of a searching investigation of the problem of bringing energy to industrial processes."[32] The "long" part of Campbell's prediction was to prove true; the "happy" part would be another story. Happy marriages require compatible partners. Dow Chemical needed power at the lowest possible price. Consumers Power, a regulated utility, needed to build a plant that it believed the Michigan Public Service Commission would not only approve but would also allow passing the cost through to utility customers. During construction, Consumers Power's managers would convince themselves that the commission would approve the plant without regard to its cost. With fundamentally different needs, the Dow Chemical–Consumers Power marriage was predestined not to be a happy marriage.

Consumers Power

Challenges of Managing New Notions

For the Dow Chemical Company, the Midland Nuclear Plant was the answer to making its Midland chemical complex profitable with low-cost energy, and it aligned with its longstanding successful efforts to continually introduce new products and new processes. In 1967 the Consumers Power Company was a staid, conservative, old-line Midwestern electric and natural gas utility. Its finances were in such good shape that in April 1967 it declared a 10 percent stock dividend and continued to pay its quarterly cash dividend on the increased number of shares. The company prided itself in being the reliable mom-and-pop stock; its 1967 annual report bragged that a large majority of its common stock, about 89 percent, was held by individual stockholders and joint accounts, often husband and wife.[1] Indeed, one of Consumers Power's old-timers, Harry Wall, senior vice president in charge of electric operations in the 1960s, was skeptical of nuclear power because he feared the company might go overboard on new notions.[2]

How did Consumers Power come to embrace this new technology? The quest for an answer leads to one man, Jim Campbell. As Campbell rose through the executive ranks at the utility, he constantly advocated for the use of nuclear power for power generation. Not only did he push for nuclear generation, he also had the utmost confidence that the company could successfully build nuclear plants. Unfortunately Campbell did not live to direct his nuclear vision for Consumers Power.

A 1933 Purdue University engineering graduate, Campbell had in 1939 been awarded an Alfred P. Sloan Fellowship for a year of graduate study at the Massachusetts Institute of Technology.[3] In 1952, just two years after he was promoted to vice president, Campbell began his campaign for nuclear power.[4] He was named a senior vice president in 1956 and then president in 1960. In the 1960s and early 1970s, he was successful in getting Consumers Power to commit to build ever larger nuclear-powered electric generation facilities. As an engineer, he had the technical background to select the newly available nuclear energy technology as the next step in a logical progression of Consumers Power's source for electrical generation.

Campbell could also have appealed to Consumers Power's board of directors to undertake an aggressive campaign to build nuclear power plants with the argument that the company needed to play catch-up with Detroit Edison, Michigan's other big utility. Through the dynamic efforts of Walker Cisler, Detroit Edison's president and later chairman of the board, the company had provided the early leadership for nuclear power generation in Michigan. In December 1951, it signed an agreement with the Atomic Energy Commission (AEC) to undertake a study of the practicability of using nuclear power for industrial purposes with the main objective of creating a large-scale reactor to produce power and fissionable materials as joint products.[5] I can imagine that as a proud and dedicated Consumers employee, Campbell could argue that his company should not appear to be the little brother of the big two Michigan utilities, lagging behind big brother Detroit Edison.

Other arguments in favor of Consumers Power's using nuclear generation in 1967 were similar to those motivating Dow Chemical: the high cost of coal and concerns about pollution from coal. In addition, ironically, Consumers Power would be able to increase its profit if it built nuclear plants instead of less expensive coal plants.

The company that grew into Consumers Power originated in Jackson, Michigan, in the 1880s, when a young miller, W. F. Foote, saw the potential for lighting streets with electricity. His preferred method of generating electricity was using dams to provide the water power to run electric generators. He and his successors expanded his original company by building some dams and acquiring others through a long series of mergers and acquisitions. By the 1930s, Consumers Power owned dams on the Au Sable, Manistee, Muskegon, Pine, Grand, and Kalamazoo Rivers in

Michigan's Lower Peninsula.[6] In 1931, at a cost of $5.29 million, the utility completed the Hardy Dam on the Muskegon River, its last and largest. This dam could generate 31 megawatts (MW), approximately enough to power twenty-three thousand to thirty-one thousand homes.[7]

But, as noted earlier, Michigan's many rivers don't have the potential to provide large quantities of energy for electrical generation because of the state's fairly flat topography. If Consumers Power wanted to continue to expand, it needed another source of fuel. From dams, it turned to coal-fired steam plants. One example is the B. C. Cobb Plant in Muskegon County. Its five units, with a total capacity of 520 MW, were built between 1948 and 1957.[8] Another example is the J. H. Campbell Plant located near Grand Rapids, the first two units of which were completed in 1962 and 1967. With the addition of the second unit, it had a 669 MW generating capacity.[9]

Whereas the first two units at the Campbell Plant cost approximately $130,000 per MW of capacity, when the Midland Plant was proposed in 1967 with a projected cost of $267 million, its cost per MW of generating capacity (1,300 MW of electricity and 300 MW of equivalent steam capacity) was $167,000.[10] Despite the higher construction cost for the nuclear facility, Consumers Power said that, because of the cost to ship coal to the Midland area, it would generate electricity for less cost than a coal-fired plant would.[11]

Another argument for not building new coal-fired generation in the late 1960s was the growing awareness of the pollution that resulted from burning coal. In response to this concern, by the late 1960s Consumers Power had committed to an $18 million three-year program to install electrostatic precipitators and other pollution control equipment at its major coal-burning plants.[12] At the same time, the company also recognized the difficulty of overcoming the emission of sulfur dioxide from coal-fired plants, which caused acid rain. Consumers Power's evaluation of this problem in its 1969 annual report was that although high stacks (smokestacks) could carry off emissions and disperse them, this was not the complete solution. Using low-sulfur coal or oil could be helpful, but the world supply of low-sulfur fuel was insufficient to make it practical.[13]

Consumers Power's report to its shareholders on February 16, 1970, had this to say about the solution to the pollution from coal-fired plants:

> A major advance in environmental protection is the construction of large nuclear plants. These clean, quiet, economical generating stations can be located near load centers, without necessarily being placed in unspoiled rural areas. They produce

no combustion products, and adherence to rigid safety standards ensures safety from radiation.[14]

But perhaps one of the most compelling reasons for a utility like Consumers Power to build nuclear power plants was the opportunity to make more money. Utilities took the position that the lower fuel costs for nuclear power would offset the higher front-end costs of building nuclear power plants and produce cheaper electricity for their customers over the lifetime of a plant. In the peculiar world of public utilities, the fact that a nuclear plant cost more to build than a coal plant of the same size allowed the utility to make more money because it increased the size of their capital investment (often referred to as their rate base). After allowing for the utility's operating cost, the state utility commission calculated the rate of return allowed on the utility's rate base. The bigger its rate base the more money a utility makes. As a result, while not affecting the rate of profit per se, a larger rate base did increase the absolute profits accruing to a company.[15] For example, if a utility was able to borrow money at 6 percent to increase its rate base by building electric generation, and the utility commission allowed a rate of return of 11 percent on the utility's rate base, it was in the utility's best interest to build expensive generation capacity.

Campbell got his first chance to launch Consumers Power into nuclear power through participation in the Power Reactor Development Company (PRDC). The early efforts of Detroit Edison and Dow Chemical had led to the formation of the Atomic Power Development Association and then the PRDC.[16] By 1954, Campbell, then a vice president, was the driving force behind Consumers Power's participation in the PRDC, which built the Fermi 1 breeder reactor south of Detroit in 1956.[17]

Thanks to Admiral Hyman Rickover's work, light water reactors replaced the Fermi 1 design. Both General Electric and Westinghouse were aggressively promoting their light water designs in the United States in the late 1950s and 1960s. In 1958 Campbell, now a senior vice president, attended a meeting conducted by Ralph Cordiner of General Electric in Woodside, California, to hear General Electric's sales pitch for its light water reactors design. The material presented convinced Campbell that the future of electric utilities lay in using these new light water reactors to generate electricity.[18]

General Electric's design was a light-water boiling water reactor (BWR). The Westinghouse design was called a light-water pressurized reactor (PWR). Although Westinghouse and General Electric were the leaders in producing light water reactors, both Babcock & Wilcox and Combustion Engineering later joined the competition.[19]

▪ ▪ ▪

Consumers Power's first entry into building nuclear reactors for its own generation was a small, 67 MW General Electric BWR at Big Rock Point. Started in 1960, this plant was located on the shore of Lake Michigan near the small town of Charlevoix. Bechtel was the prime contractor, and it built the reactor in twenty-nine months under a fixed price contract at a cost of $27.7 million, with construction completed in 1962.[20] Even though for a whole year of operations the Big Rock Point Plant was plagued by serious vibrations in the neutron-absorbing safety shield within the reactor, Campbell felt it had given the utility the construction and operating experience needed to forge ahead with nuclear power.[21]

Campbell described the Big Rock Point plant as being a research and development project not capable of producing electricity as cheaply as the company's coal generating plants. At a news briefing in 1960 at the site of the Big Rock Point plant he said, "We feel that our company, like any other large business enterprise has the responsibility of looking ahead; anyone who expects the world to stand still is going to be disappointed."[22] But in 1960, Campbell predicted that nuclear power would be competitive with coal by 1968.[23]

His faith in the future of nuclear power aligned with the general consensus of the 1960s. Although the Shippingport Plant, the world's first full-scale atomic electric power plant devoted exclusively to peacetime uses, was a success as a proof of concept, its generating cost was several times that of a fossil fuel plant. It produced electricity at a cost of 64 mills per kilowatt hour (6.4 cents/kWh) at a time when conventional coal-fired plants were producing at 6 mills per kilowatt hour (0.6 cents/kWh).[24]

Five years after Shippingport became operational, the AEC sent the White House a study titled *Civilian Nuclear Power—A Report to the President—1962*. This report highlighted the decline in the cost of nuclear electricity generation since the first plant had been built and concluded that nuclear power was on the threshold of economic competitiveness. It could soon be made competitive in areas consuming a significant fraction of the nation's electrical energy.[25] Just a year later, an apparent breakthrough in nuclear electric generation costs came when Jersey Central Power & Light Company announced the purchase of a 515 MW BWR from General Electric. The cost of generating electricity from the utility's planned Oyster Creek Plant was projected to be lower than the cost of generating from any other type of plant that could be installed at the same location, appearing to confirm the report's prediction.[26]

It is significant that in the period from 1967 to the early 1970s Campbell, Consumers Power, and the utility industry in general did not seem to comprehend that the reported cost for early nuclear plants was not the true cost. A 1977 study concluded that General Electric and Westinghouse had actually lost $875 million to $1 billion in total on the thirteen nuclear power plants they had constructed on turnkey contracts.[27] These early turnkey projects turned out to be what retail merchants call "loss leaders," a product sold at or below cost to lure buyers into the store, where they will probably purchase something else.[28] These loss leader contracts produced results. Utilities all over America were ordering nuclear power plants to be built at an ever increasing rate in 1967. Whereas in all of 1966 only twenty-one reactors had been ordered, during just the first six months of 1967 a total of twenty-two electric utilities announced plans for twenty-three nuclear power plants with a total capacity of about 18,325 MW.[29]

In 1966 Consumers Power announced that it would build its Palisades Plant, a 700 MW facility on the shores of Lake Michigan in Van Buren County. In the press release for the project in December 1966, Campbell said, "We already demonstrated the practical value of a nuclear reactor at our Big Rock Point plant, near Charlevoix. We have proven there that a nuclear-fueled plant can be applied to our system on a large scale, economically and efficiently. So it just makes good sense to use nuclear fuels as well as coal in meeting Michigan's needs for electric energy."[30]

In 1966 not only did Campbell think that Consumers Power had learned how to run nuclear plants, he also believed that the company knew how to build nuclear plants and didn't need to obtain turnkey contracts. In an article in *Electric World*, Campbell said, "Our company now feels confident that it can apply standard procurement and construction concepts . . . The opposite concept—turn-key—inevitably must and does include extra charges for contingencies which may never occur. If a contingency presents itself, we will have to face it; but we see no reason to pay a lump sum for a whole series of contingency provisions."[31]

Consumers Power was not alone in deciding to build nuclear facilities without turnkey contracts. Between December 1963 and April 1967, while General Electric and Westinghouse sold thirteen turnkey plants, utilities bought twenty-seven non-turnkey plants.[32]

The Palisades Plant was dramatically bigger and had a completely different design from Consumers Power's first nuclear plant at Big Rock Point. Palisades Plant's PWR was designed to generate approximately 700 MW of electricity, almost ten times the capacity of Big Rock Point's BWR.[33] The Palisades Plant selected Combustion Engineering to manufacture the PWR reactor and steam supply system. The selection was consistent with Consumers Power's view that it had the competence to pick components for a nuclear power plant from a variety of suppliers. But the Palisades Plant steam system and nuclear reactor were the first ones that Combustion Engineering would build and experienced numerous problems when it started operating.

Consumers Power also switched reactor manufacturers between the two projects, using General Electric at Big Rock Point and then Combustion Engineering at the Palisades Plant. In 1974 Russell Youngdahl, senior vice president for electricity, explained that Consumers Power's choice of Combustion Engineering was meant to set an example to the industry. Although he had not been in charge at the time the selection was made, he said, "There was some concern within our company that entry into the nuclear business was getting too restrained, and we'd have only two vendors: GE [General Electric] and Westinghouse."[34]

Consumers Power did not have a turnkey contract for the building of the Palisades Plant, which was under construction from 1966 to 1971 with a reported completed cost of $188 million (when announced in 1966, its estimated cost was $100 million).[35] Not known at the time the Midland Plant was announced in 1967 was that Palisades Plant would eventually encounter a series of problems that would result in the utility's filing a suit in 1974 for $300 million against Bechtel Corporation, Bechtel Company, Combustion Engineering, and two other contractors for faulty equipment that resulted in shutdowns of the plant.[36]

▪ ▪ ▪

As initially planned, the Midland Plant was intended to have a capacity of 1,300 MW of electric generating capacity produced by side-by-side reactors. The design was also to have the additional complexity of the capacity to produce four million pounds of steam an hour for Dow Chemical. The steam production would be the equivalent of approximately 300 MW of electric production, and therefore the Midland Plant would be the equivalent of a 1,600 MW facility.[37] As it had for the Palisades Plant, Consumers Power passed over the most experienced manufacturers, General Electric and Westinghouse. As the utility dramatically increased the generation capacity of the Midland Plant compared with its two earlier nuclear plants, Consumers Power also selected a new supplier for steam systems and reactors—Babcock & Wilcox.

Consumers Power's plans to build new nuclear generating facilities did not end with the Midland Plant. In 1971 the company announced studies for the possible construction of a 2,300 MW plant at an estimated cost of $1.2 billion on the shore of Lake Huron at Quanicassee near Bay City, twenty-seven miles east of the Midland Plant.[38] In late 1972 it reported it had ordered two Westinghouse PWRs that were tentatively destined for a site at Quanicassee.[39] In November 1973 Consumers Power filed a preliminary application with the AEC for approval to build the Quanicassee plant.[40] Had both the Midland Plant and Quanicassee been completed, 45 percent of Consumers Power's generating capacity would have come from nuclear plants in 1980.[41]

With its announcement about the Quanicassee plant in 1971, Consumers Power predicted that the peak demand on its system in 1980 would be twice the peak demand in 1970. The press release included a comment by Youngdahl that projected electrical energy demands for Michigan in the 1980s showed that a number of large central generating plants such as the Quanicassee unit would be needed.[42]

While Consumers Power proceeded with its aggressive nuclear plans, Michigan's other utility company, Detroit Edison, announced only one nuclear plant in 1968, its Fermi 2 plant (1,100 MW) located on the shore of Lake Erie in southeastern Michigan.[43] Detroit Edison selected equipment for a BWR from General Electric.[44] By 1969 it appeared that little brother had a plan to beat big brother in the race to build nuclear energy facilities.

Amid the rapid build-up, Campbell's role in guiding Consumers Power's nuclear program was cut short. Then president of the company, he died from a heart attack on January 24, 1972, at the age of sixty-one.[45] Rather than promote one of their vice presidents, the board requested A. H. Aymond, chairman of the board, to assume the additional title of president. A lawyer, Aymond had worked his way up through the ranks in a series of legal jobs.[46] He would hold the title of president until 1975.

By 1973 Consumers Power's old guard agreed with Campbell that the company was "not going overboard with new notions" with its aggressive building program of nuclear plants. Harry Wall, who had been named vice chairman of Consumers Power's board of directors in July 1972, said about nuclear power,

> There is nothing else in the picture. The coal supply is becoming critical. Even coal that's full of sulfur is hard to get. Oil and gas are limited. There isn't enough

> of either available to carry a system like ours. The fuel cell is not developed to the point where it could be a central-station source (as opposed to individual use, as in a space capsule), and solar energy isn't developed at all to my knowledge. I can't see any other fuel at the moment except nuclear that can take care of large supplies.[47]

As Consumers Power made the move to nuclear power, it also dramatically increased the size of each new nuclear plant: in 1962 at Big Rock Point, 67 MW (one reactor); in 1966 at Palisades, 700 MW (one reactor); in 1967 at Midland, 1,600 MW (two reactors to produce 1,300 MW electricity plus steam for Dow Chemical); and in 1971 at Quanicassee, 2,300 MW (two reactors). In addition, it selected different equipment manufacturers for each of its nuclear plants: in 1962 at Big Rock Point, General Electric, BWR; in 1967 at Palisades, Combustion Engineering, PWR; in 1968 at Midland, Babcock & Wilcox, PWRS; and in 1971 at Quanicassee, Westinghouse, PWRS.

Looking at the history of Consumers Power's rush to build nuclear power plants, it is difficult to see how the leadership thought the completion of the 67 MW plant at Big Rock Point gave them the confidence to forge ahead with plants ten to thirty times that size with no two plants using the same equipment manufacturers. But implicit in the company's 1967 decision to build the Midland Plant was the new notion that it could successfully build large nuclear power plants. Lacking the necessary staff to manage all of the industrial contractors necessary for building the Midland Plant, Consumers Power selected Bechtel to be the architect, engineer, and construction supervisor. Had Aymond adopted the management organization and supervision style of Rickover, this story might have ended quite differently.

CHAPTER 7

Let Bechtel Do It

Just Make It a Megaproject

As a journalist once said of Bechtel Group, "It's the stuff of fiction thrillers, a giant company, controlled by one family from a skyscraper overlooking San Francisco Bay employing former high government officials operating around the world with rumored links to the CIA."[1] But this company wasn't fiction. Composed of three companies—Bechtel Corporation, Bechtel Inc., and Bechtel Power Corporation—it was a very big, very well-connected company, and very private company located in San Francisco. Although the Bechtel Group was most frequently referred to by one of the three company names, it liked to refer to its activities as simply being done by Bechtel since it was constantly reorganizing itself to take advantage of new opportunities. During the time of our story some of these new major Bechtel entities were the Saudi Arabian Bechtel Co.; Bougainville Copper Project in Papua, New Guinea; and Bechtel Energy Group.

The founder of what would become the Bechtel empire, Warren Bechtel, had started out in 1898 building railroads in the western United States and developed a company that took the lead in constructing the Hoover Dam in the 1930s. When he died in 1933, in the middle of that project, his son Stephen D. Bechtel Sr. took over running the company.[2] After his retirement in 1960, the lineage of top management later included Stephen D. (Steve) Bechtel Jr. and then his son Riley P. Bechtel in 1989.

In 1978 the three Bechtel companies had revenues that totaled $6.4 billion, up from $4.5 billion two years earlier. Profits in 1976 were $66.5 million (in 1979 when this information was published, estimated profits for 1977 and 1978 had not been disclosed). *Engineering News-Record*, a trade publication, put the group's pretax profits at 42 percent of stockholder equity for 1976, compared to average construction industry profits at about 30 percent of stockholder equity.[3] Whereas Consumers Power Company and Dow Chemical Company were publicly traded companies based in the flyover country of the American Midwest, Bechtel's projects evoke the lines from the movie *Mary Poppins* (1964) when young Michael Banks is being advised to invest in the Fidelity Fiduciary Bank and is told that it was involved with "Railways through Africa / Dams across the Nile / Fleets of ocean greyhounds / Majestic self-amortizing canals / Plantations of ripening tea."[4]

Bechtel played a significant role in the saga of the Midland Nuclear Plant from its conception in 1966 to 1984 when construction stopped. As noted earlier, Consumers Power and Dow Chemical initially engaged Bechtel in 1966 to do an engineering and economic feasibility study of whether, given the location and purpose of the facility, a dual-purpose facility (steam for Dow Chemical and electricity for all of Consumers Power's customers) would be economically desirable for both parties.[5] As a part of this work, Bechtel also prepared conceptual layouts for plants of various sizes, estimated the capital requirements, and developed the plant specifications.[6] Consumers Power had the benefit of this study when it announced in December 1967 that it would construct the Midland Plant for $267 million and start operating the first reactor in 1974. Bechtel's role continued after this initial study as Consumers Power retained the firm as the constructor, engineer, and architect.[7]

The decision seemed like a good one. Bechtel prided itself on a long history of accomplishing megaprojects, including the San Francisco–Oakland Bay Bridge in 1933, Liberty ships in 1941, Trans-Arabian Pipeline in 1947, Suncor Athabasca Tar Sands Project in 1962, San Francisco's Bay Area Rapid Transit System in 1964, and Washington D.C.'s Metro transit system in 1971.[8]

When the company took on the Midland Plant, Steve Bechtel Jr. was president and chief executive of the Bechtel Group. When he visited the Midland Plant on June 30, 1981, he was both chairman and chief executive. After that visit, Bechtel site manager Leo Davis said, "Mr. Bechtel was very complimentary about the project and expressed his appreciation to those involved."[9]

But unlike the scene in *Mary Poppins* where young Michael Banks is encouraged to invest his money in the Dawes, Tomes, Mousley, Grubbs, Fidelity Fiduciary Bank, the public couldn't buy shares in Bechtel, as it was a private company. In 1978 a *Mother Jones* article reported that all the Bechtel shareholders (about fifty-six) were either company vice presidents or their wives.[10] All of the rules and regulations of the U.S. Securities and Exchange Commission for public companies did not apply to private companies throughout the time the Midland Plant was under construction. This meant that Bechtel had no requirements to file detailed financial reports with the government, to disclose payments to foreign governments or individuals, nor to make public the details of its financing arrangements.[11]

However, Bechtel's extremely limited obligations for reporting to government entities does not mean that the company did not take an interest in the government. Dating back to the days of building Liberty ships during World War II, Bechtel took a strong interest in being well connected in Washington, D.C. In fact, Stephen Bechtel Sr.'s friend and business partner, John A. McCone, after making millions with Bechtel during the war, moved into the highest levels of the U.S. government, in 1958 becoming chairman of the Atomic Energy Commission (AEC). During McCone's tenure, Stephen Bechtel Sr. was a familiar face at the Eisenhower White House.[12]

In 1958 Bechtel hired W. Kenneth Davis, head of the AEC reactor program, to open the company's new nuclear division. Davis said, "Nuclear power was a mechanism for getting Bechtel into the power plant business."[13] He was an advocate of building nuclear power plants as close as possible to the consumers, stating that plants on the outskirts of New York City "will not bring undue safety hazards to plant workers or the public."[14]

Bechtel's association with nuclear power began in 1949, almost a decade before Davis joined the firm, with the construction for the AEC of the experimental fast breeder reactor facility EBR-1 in Idaho. By the time Bechtel's study of the Midland Plant came out in 1967, its standard press release recited a list of additional nuclear projects: "first Dresden unit in Illinois, Humboldt Bay in California, Big Rock Point, Nuclear Fuel Services fuel reprocessing plant in New York, Hallam Plant in Nebraska." Also noted were heavy-water reactors in Canada and Sweden as well as the advanced gas-cooled reactor of the United Kingdom Atomic Energy Authority.[15]

But "getting into the power plant business," as Davis said, meant that the company was not interested in making reactor and stream supply systems like Westinghouse and General Electric. Its goal was to be the architect, engineer, and construction supervisor of everything that went into building a nuclear plant—this was the megaproject in the rush to build ever larger nuclear plants in the late 1960s and early 1970s. In 1965 Bechtel was successful in obtaining the contract for the engineering and installation of Florida Power & Light's Turkey Point Three and Turkey Point Four Plants.[16] According to an engineer who worked for the company then, this job was done on a fixed-price contract.[17] Landing this job put the company in an excellent place to compete for the contracts to build the Palisades and Midland Plants.

Having established its preeminence in its successful effort to win the Turkey Creek work, Bechtel obtained contracts in the late 1960s for the Palisades and the Midland Plants. However, the engineering firm did not enter into fixed-price contracts for these projects. In addition to its preference for enormous projects, Bechtel was famous for its cost-plus contracts. Along with John Alex McCone, Stephen Bechtel Sr. has been credited with inventing in the 1930s the "cost-plus" model in which the government guaranteed contractors all costs of production plus a built-in profit of 10 percent.[18] Not having a fixed-price contract with Bechtel for the Midland Plant did not matter to Consumers Power, which Jim Campbell had earlier decided the utility didn't need.[19]

As the late 1970s and early 1980s progressed, Bechtel hired more and more people to work on the Midland Plant. One Bechtel employee described the work there as a cost-plus contract in which the less work gets done, the more money you make, with this philosophy extending all the way down the ranks.[20] In 1984 the *Lansing State Journal* reported that Bechtel had billed Consumers Power $1.9 billion for its work on the Midland Plant.[21] This was for a project that in 1967 had had a total estimated completion cost of $267 million. The actual cost, which soared from $267 million to over $4 billion for an uncompleted project, certainly moved it into the category of a megaproject.

1968

Forces Not under Your Control Matter

Immediately following the December 14, 1967, announcement of the Midland Plant, the Consumers Power Company and Dow Chemical Company began spreading the good news to the local community. On January 9, at the Midland Community Center, executives from the two companies outlined the implications and advantages of the project to 120 members of the Midland Chamber of Commerce at a monthly before-work pep rally they called their "Eye-Opener" meeting. The three executives were William Kessler, Consumers Power project manager for the plant; Harold Bosscher, general manager of the Midland division of Dow Chemical; and James Albertson, Consumers Power area manager.[1]

In the Eye-Opener, Bosscher enthusiastically described how different groups would thrive as a result of the construction of the Midland Plant. Production workers' jobs would be more secure because costs in Midland would now be more competitive with Gulf Coast costs. Construction workers would benefit because Dow Chemical's Midland division would consider renovation and replacement of equipment and facilities. Dow Chemical's three thousand Midland scientists would have greater opportunities because with cheaper energy they would be able to construct more pilot plants. Citizens in Midland would breathe cleaner air once Dow Chemical stopped burning coal. Kessler's focus was on jobs. He said construction of the plant would begin in the winter of 1969, and during the construction phase, between two hundred and five hundred people would be employed and that the first of the two reactors was scheduled for completion in 1974.

At the conclusion of their presentation, Chamber of Commerce president W. Victor Bielinski presented Bosscher and Albertson certificates of appreciation for their "major effort and investment which will permit continued and exciting growth for the Midland area." But despite the happy themes emphasized at the Eye Opener, significant issues not mentioned turned out to affect the plant's construction in ways that neither Consumers Power nor Dow Chemical could control. These issues included loss of public trust in the government experts, inflation driven by the Vietnam War, future ramifications of equipment supplier choice, and growing public concern about the safety of nuclear power. The ultimate success of the Midland Plant depended on the ability to manage not only all of these external items, but also others that would later present themselves.

The Midland Plant provided a welcome bit of good news in 1968. It was an antidote to the bad news about the Vietnam War, which was day after day the page one story in the *Midland Daily News*. This war took place before America turned over its military obligations to an all-volunteer force. During the Vietnam conflict, every American male above the age eighteen could be compelled to serve in the military, unless he qualified for an exemption or deferment from military service.[2] "Many Yanks Being pressed into Second Vietnam Tours" was front-page headline in the *Midland Daily News* for October 14, 1968. More than 8,615,000 draft-age men served in the U.S. armed forces during the war.[3]

As the situation in Vietnam deteriorated, President Lyndon Johnson decided again and again to spend more money and send more troops. Although he did so regularly from 1963 to 1968, this was not Johnson's war. John F. Kennedy had proclaimed in his inaugural address in 1961, "Let every nation know, whether it wishes us well or ill, that we shall pay any price, bear any burden, meet any hardship, support any friend, oppose any foe to assure the survival and the success of liberty." He meant that the United States would stop Communism from spreading anywhere, including Vietnam. When Johnson assumed the presidency in 1963 after Kennedy was assassinated, he inherited a war he did not like and did not understand. Indeed, he commented to an aide, "I feel just like I grabbed a big juicy worm with a right sharp hook in the middle of it."[4]

Johnson saw improving the lives of all Americans with a war on poverty as his mandate. He led the struggle with more than two hundred important pieces of legislation, including the Civil Rights Act of 1964, the Voting Rights Act of 1965, federal aid to education, Head Start, Medicare, and a whole series of bills aimed at ending poverty in America.

But by 1968 the Vietnam War was taking a toll on Americans' trust in experts and the government. There were over five hundred thousand service members in Vietnam, and war expenditures were growing monthly in that year. On June 14, the *Midland Daily News* reported that from 1965 until the present, 24,477 American men had been lost in combat in South Vietnam. According to a Gallup poll, 61 percent of Americans supported the Vietnam War in 1965, but by the spring of 1967 their support fell permanently below 50 percent. By the last half of 1968 support was between 35 percent and 40 percent.[5]

Two hundred people with the organization Clergy and Laymen Concerned about Vietnam came to Midland for Dow Chemical's annual meeting in 1968.[6] They protested the war and the fact that Dow Chemical was selling napalm to the government for use in Vietnam. Leaders of the protest movement, like actress Jane Fonda, intensified their anti-war activity after 1968, but 1968 was the point that many Americans were beginning to question what their leaders had told them about Vietnam, why Americans were dying, and why hundreds of millions of dollars were being spent on a war in Asia.

▪ ▪ ▪

How did the war in Vietnam affect the construction of a nuclear plant in Midland, Michigan? In a word: inflation.[7] Because of the unpopularity of the war, President Johnson was reluctant to raise taxes or cut his social programs to deal effectively with the inflation generated by the demand for men and materials.[8] Year-to-year changes in the Consumer Price Index, which had been increasing at an annual percentage rate of between 1.1 percent and 1.6 percent from 1960 to 1965, changed dramatically as the Vietnam war wore on, rising to 3 percent in 1966, falling slightly to 2.8 percent in 1967, shooting up to 4.3 percent in 1968, going up again to 5.5 percent in 1969, and rising in 1970 to a grim 5.8 percent.[9]

This made it necessary to pay more interest to borrow money to cover the cost of the most recent rate of inflation. In 1967 Consumers Power sold bonds at an interest cost of 5.9 percent to the company, but by 1970 its bonds were sold at a rate of 8.6 percent.[10] Inflation also caused the cost of labor, building materials, and services to go up. The *Engineering News-Record* historical record shows that building costs, which had escalated at 3.6 percent and 3.4 percent for 1966 and

1967, increased by 7.1 percent and 9.8 percent for 1968 and 1969.[11] So in 1969 the estimated cost of the Midland Plant had grown from the original 1967 estimate of $267 million to $375 million.[12]

In Consumers Power's 1968 annual report, management identified inflation as the most serious problem facing the company and all other investor-owned utilities:

> A 6% or 6 1/2% rate of return on a utility's investment in facilities is not adequate under economic conditions existing today. As the cost of capital increases, so must the rate of return the utility is allowed to earn on its investment . . . Sharply increased interest costs on the large amounts of money the Company must borrow for its expansion program and tremendous increases in construction costs continue to be major sources of economic pressure on the Company.[13]

The year 1968 gave Consumers Power a preview of the damage inflation could cause not only in the cost to borrow money to fund the Midland Plant, but also in rapidly rising construction costs. The fact that Consumers Power did not have a fixed-price contract left it responsible for finding the money to finish the plant, whatever the cost.

In May 1968 a small item about the Midland Plant was tucked into page 3 of the *Midland Daily News* next to a story about a local man who had been injured in Vietnam.[14] This seemingly insignificant article reported on the selection of major equipment suppliers for the Midland Plant, noting that Consumers Power had selected Babcock & Wilcox to provide the light-water pressurized reactors and the steam supply systems.[15] The good news was, unlike with the Palisades Plant, the Consumers Power order for the Midland Plant was not Babcock & Wilcox's first for this kind of equipment. However, the problem in both cases was that Consumers Power had avoided ordering from the most experienced manufacturers of nuclear reactors, Westinghouse and General Electric.

According to the *Midland Daily News*, the other eight Babcock & Wilcox units like the ones that would be provided to Midland had been ordered by Jersey Central Power & Light, Duke Power Company (which ordered three units), Metropolitan Edison, Florida Power Corporation, Arkansas Power & Light, and the Sacramento Municipal Utility District. Although Consumers Power could not have known it in 1968, its selection of Babcock & Wilcox equipment would later have great significance: a similar unit would be installed at Three Mile Island and operated by Metropolitan Edison. Changes mandated for the Midland Plant and a major loss of confidence in the safety of nuclear power after the accident at Three Mile Island would have serious consequences for Consumers Power.

Finally, one of the most significant events of 1968 for the Midland Plant was Mary Sinclair, who lived about five miles from its proposed location, started to speak and write about the public safety, risks, and pollution potential of the proposed nuclear facility. It quickly became apparent that Sinclair did not agree with what the executives at either Consumers Power or Dow Chemical had to say about the safety of the Midland Plant. In Midland in 1968, it was preposterous that a *woman* would question the actions of Dow Chemical, the third largest chemical company in America; Consumers Power, one of the ten largest combined utilities in America; and Bechtel, one of the most famous construction companies in America.

CHAPTER 9

Mary Sinclair

No Typical 1960s Woman

Until 1979 press reports on the Midland Plant almost always referred to Mary Sinclair as Mrs. William D. Sinclair and included her street address. Referring to a woman by her married name was was standard in the 1960s and early 1970s; giving her street address as part of her identification seemed to indicate that she was busy in her kitchen baking cookies for her children. It may have also been a carry-over from the society pages, where women were more likely to be featured than in news articles. But Sinclair was not destined to appear only in reports about PTA meetings and the Midland Symphony League.

In January 1968, Sinclair was a forty-eight-year-old woman married to a successful local attorney, mother of five, member of the Family Service Board, parishioner at Blessed Sacrament Catholic Church, and supporter of the Midland Symphony. Her public advocacy prior to 1968 had involved working on the statewide establishment of temporary mental health facilities to provide a place for afflicted people to go other than to asylums and jails. She also received the Distinguished Service Award of the Michigan Congress of PTAs in recognition for articles she had written suggesting improvements in and greater understanding of the education system.[1] She could best be described as a respected citizen of Midland.

Sinclair's life changed forever in 1969 as she began to speak and write about

FIGURE 7. Circa 1972–1978, Mary Sinclair in her home office. (Courtesy Bentley Historical Library, University of Michigan, "Mary Sinclair," HS1426, Mary Sinclair Papers)

her concerns with the safety of the Midland Plant. In a letter to the *Midland Daily News* in June of that year, she wrote, "The environment today is not just unlimited and physical—to be overcome—as it was in the past. Today the environment is limited and social. It must be cared for—even loved . . . let us have nuclear energy provided—but with maximum pollution controls and safety factors secure at the outset."[2] In late 1968 and 1969 she wrote and distributed in Midland copies of a booklet, *Nuclear Power & Public Concern*, in which she pointed out the problems with nuclear power.[3] Then on December 18, 1969, she served on a panel on a local television program, *Nuclear Power, Benefits and Risks*, during which she pointed out the problems with the Midland Plant. The *Midland Daily News* quoted her comments following the television program: "No chances should be taken with waste material in the environment that could have genetic effects on humans."[4] In a *Midland Daily News* editorial the paper's editor, Norman (Slim) Rumple, wrote that the show's panel had apparently been chosen to inject controversy into the Midland area.[5]

Sinclair's response to his editorial, published as a letter to the editor, read,

> Your editorial of Saturday, December 20, implied that questioning the present course of nuclear power development in Midland was injecting irresponsible

> controversy into the community. I hope that you as a journalist are not suggesting that matters of public concern are not proper subjects for public discussion. Since your editorial was filled with questionable statements, I hope you will open this column to a free discussion of the issue by the community.[6]

By the end of 1969, because of her public questioning of the safety of the Midland Plant, most Midlanders described her as a job killer, anti–Dow Chemical Company, and a denier of the necessary march of progress. Neighbors began avoiding her at the supermarket and Midland Symphony performances, and her children were insulted at school by other students and even by teachers. Years later her daughter Rosemary told her that she would run to the bathroom at school sometimes so that people could not see her cry.[7] People threw garbage on her lawn, and she believed that her telephone was tapped.[8]

Sinclair's husband, William, had been chairman of the Midland County Republicans, had a thriving law practice in Midland, and was selected in 1968 to deliver the commencement address at Midland High School, which had named him a distinguished alumnus.[9] As her activities opposing the Midland Plant continued, his law partner left him, remarking that "lawyers' wives shouldn't get into controversial issues." Several of the most outspoken advocates of the nuclear plant even visited local bankers, real estate agents, and merchants, persuading them to take their business away from him.[10] In a small town like Midland, the *Midland Daily News* editor, Slim Rumple, was not a stranger to the Sinclair family. Sinclair's son, Peter, said that when the Sinclairs had a cottage on the Tittabawassee River, Rumple used to come by in his boat and hang out with William, but after a while Slim did not come by anymore.[11]

▪ ▪ ▪

Raised by strict Austrian immigrant parents on the Minnesota Iron Range, Mary Jean Palcich graduated first in her high school class and went on to receive a bachelor of arts degree in English and chemistry from the College of St. Catherine in St. Paul, Minnesota. While her husband was in law school at Georgetown University, she was a technical writer and editor for the Library of Congress in Washington, D.C., where she handled classified material for the Atomic Energy Commission (AEC).[12] Terry Miller (one of the founding members of the environmental group, Lone Tree Council) recalled how Sinclair had told him that while her job at the Library of Congress was to edit articles on nuclear energy written by government employees, she couldn't help noticing that the dangers of nuclear energy were always played down.[13]

After William completed law school, the Sinclairs moved to Midland, his hometown. Once there, Sinclair worked for Dow Corning as a freelance technical writer and for Dow Chemical providing editorial assistance in the publication of *Styrene: Its Polymers, Copolymers, and Derivatives* by Dr. Ray H. Boundy and Dr. Raymond F. Boyer.[14] In addition, she was reading technical articles on nuclear energy that made clear to her that nuclear scientists were concerned about the problems of genetic damage caused by radiation and the concentration of radioactive materials once they escaped into the environment.[15] Sinclair later singled out one of these, a 1967 *Scientific American* article by George M. Woodwell, that had influenced her thinking most.[16] This piece explained that atmospheric tests of nuclear weapons had distributed radioactive material over large distances, that radioactivity could become concentrated in the food chain, and that it had taken ten years for thyroid abnormalities to appear in children who had been exposed to radiation from the United States' atmospheric test of a 15 megaton hydrogen bomb on March 1, 1954.[17]

Sinclair did not gain her understanding of the hazards of nuclear energy completely from her reading, though. In the summer of 1968, she attended a radiation hazards seminar at the Institute on Man in Science in Albany, New York, and came away with the firm conviction that she was correct in her concerns about the safety of nuclear power.[18] Returning home, she made some copies of her booklet, *Nuclear Power & Public Concern.* On the cover she noted that it contained questions that the public was entitled to have answered "*before* the massive build-up of nuclear plants now being planned and constructed become operative." Pages five through fifty-six listed thirteen topics of concern about nuclear power with short explanations. After she had distributed copies to her Midland friends, Dorothy Dow Arbury—eldest daughter of Herbert Henry Dow—came to her home and provided a donation to print 2,500 copies. The United Auto Workers later paid to print copies of a revised version in 1971.[19]

But all her knowledge, scholarship, and experience were not enough to change minds, and people's impressions of her were deep and lasting. In a 2016 interview, Sally Moss, a member of Blessed Sacrament Catholic Church in the 1960s and 1970s, recalled when her neighbor began speaking out about the Midland Plant, "I remember her leaving church and walking to her car alone. Nobody spoke to her or walked with her."[20] Sinclair said in a 1976 interview, "Well, during the [Speak Up Now] rally hostility was intense. People who were church goers, even leaders, were dealing in this community hatred too. And this, in the city of churches, you know."[21] About church she said, "One Sunday as I was getting the children ready for church, I started crying. I just couldn't go. It was too much." She did not go back to church for a long time. "For myself, I have to work things out all over again. I am sorry, really sorry, that I can no longer seem to find it in church. It [the church] was there for me at St. Catherine's, it's been there for all my life. Perhaps it will be there for me again."[22]

I mentioned in a lecture I gave to the Midland County Historical Society the hatred that some people had for Sinclair because she opposed the Midland Plant. Local lawyer, Bill Barker, who heard the talk, emailed me:

> I knew Bill and Mary Sinclair well from my arrival in Midland in 1956, eleven years before the nuclear plant came up. Bill's brother Dick shared an office with me for some time starting when I first arrived in Midland, and the small group of Midland lawyers were close. Having been friends, the dealing with the nuclear activist Mary was a very difficult thing for me . . . I think the reason for the "hatred" you mentioned was that in the general perception Mary was an oddball acting on prejudice, for lack of a better term, that she was completely mistaken in her opposition which was contrary to the science, the economics, the excitement of being on the cutting edge of the future, the facts, as we all then knew them for certain . . . We thought at the start of her opposition that she was wrong, a nut, acting for no valid reason than unjustified emotion which was endangering a beneficial project which meant a lot to Midland and Dow. We always felt sorry for her husband Bill (a truly grand guy) and their children. But we actually physically avoided her when near her, not quite shunning, more like revulsion. This perception carried over as things started to change. And emotionally still I can't help fleetingly involuntarily feeling angry when I think of Mary, though things changed and she finally did start arguing from her perception of facts from the extensive research she did which you mentioned.

One person who did support Sinclair was Dr. William Porter at the University of Michigan. She gathered up the writing she had done in the late 1960s and asked for Porter's help in developing a program that would improve her success in communicating real information about nuclear power to the public. Porter developed a graduate program to teach her how to track the way technical information made its way into the mass media.[23] One of her courses was "Special Readings in Nuclear Engineering," which she took in the nuclear engineering department under the direction of Dr. William Kerr. She used this course as preparation for her testimony on the need for better funding for nuclear fusion before the U.S. Senate and House Appropriations Committee on May 19, 1971. The gist of her testimony was that nuclear fusion would be less harmful to the environment than either fossil-fueled plants or fission reactors. She said, "Fusion development is potentially the answer to solving the environmental problems of power production."[24]

In the 1970s Sinclair's views changed from concern about the safety of power plants utilizing nuclear fission to opposition to their even being built.[25] In March and April 1970 she joined anti-nuclear teach-in events at Saginaw Valley College, the University of Michigan, and the University of Minnesota.[26] She testified at AEC hearings held on the granting of an operating license for the Consumers Power's Palisades Plant in June and before the water permit hearing for the Midland Nuclear Plant in August.[27] She also continued her efforts in Midland, speaking at the Midland Republicans Breakfast in November.[28]

Her leadership efforts also included the formation of the Saginaw Valley Nuclear Study Group on October 7, 1970.[29] This organization provided the springboard for an organized critique of the safety of the Midland Plant. The choice of the name was logical, but it also served to elevate the issue from a local one to a regional one. The plant was sited on the banks of the Tittabawassee River immediately across the river from the Dow Chemical complex. Four miles past Midland the Tittabawassee River flows past Mapleton and then continues seventeen miles until it flows into the Saginaw River south of the city of Saginaw. The Saginaw River then flows another eighteen miles through the cities of Saginaw and Bay City and then into Saginaw Bay, an arm of Lake Huron. Sinclair pointed out the problem of locating nuclear plants on waterways in her 1969 pamphlet, *Nuclear Power & Public Concern*, referring to a report that said a leak in a radioactive waste storage tank at the Hanford, Washington, facility could possibly go undetected and render the nearby Columbia River deadly (see appendix 1).

Thanks to Sinclair's leadership, five groups joined the Saginaw Valley Nuclear Study Group to form the "Saginaw Intervenors": the United Auto Workers International, the Mackinac chapter of the Sierra Club (whose national organization had not expressed a position regarding nuclear power in 1970), the West Michigan Environmental Action Council, the Citizens Committee for Environmental Protection of Michigan, and the University of Michigan Environmental Law Society.[30] The Saginaw Intervenors decided to intervene in the AEC hearings for a construction permit for the Midland Plant just thirty days prior to the filing deadline. Dr. Robert

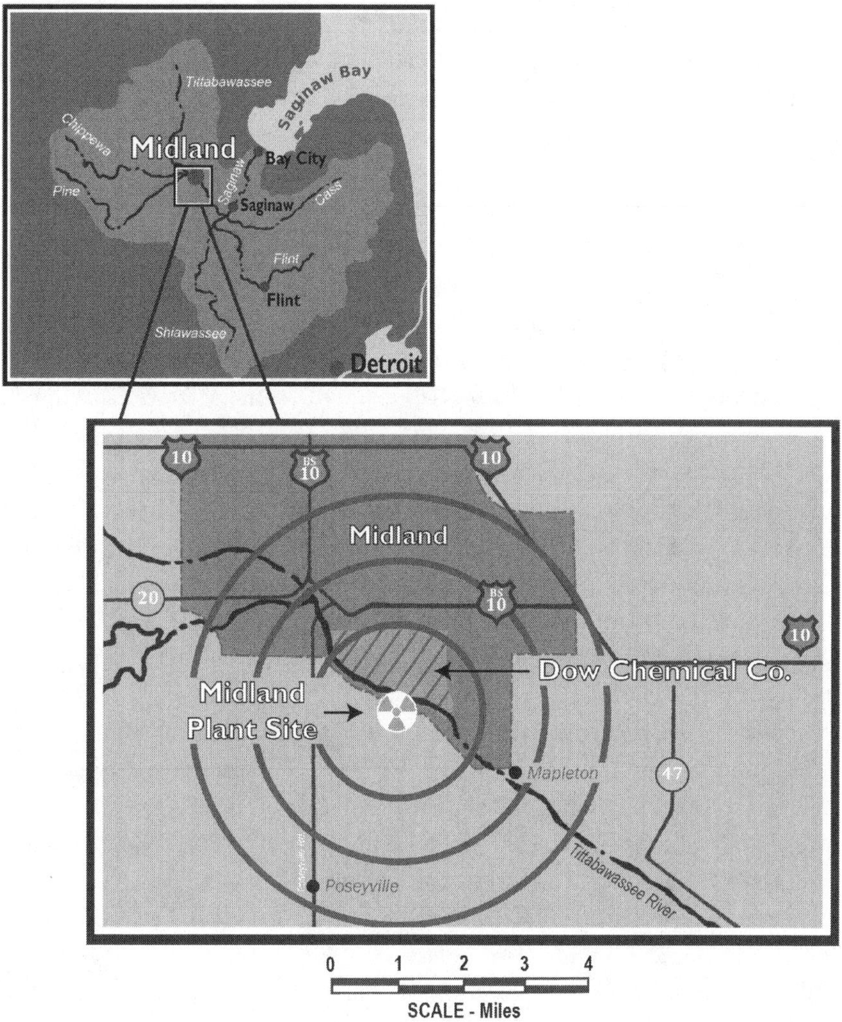

MAP 2. Location of the Midland Plant on the Tittabawassee River floodplain directly across from the Dow Chemical plant where twelve thousand people worked and within five miles of where approximately thirty-five thousand people lived. The location provided a rallying point for those who opposed the Midland Plant. Location boundaries for the city of Midland are from 1967. (Map modified by author from Consumers Power Company, *Final Environmental Statement: Related to the Operation of Units 1 and 2, Docket Nos. 50-329 and 50-330, Consumers Power Company*, vol. 1 [Washington, D.C.: U.S. Nuclear Regulatory Commission, July 1982], figure 4.1, 4–20.)

Asperger, a member of the group, said that to beat the deadline Sinclair had to fly to Washington, D.C., and deliver the necessary papers by hand.[31]

Filing the papers was only the first step in participating in the hearings. The Saginaw Intervenors needed a lawyer. Sinclair helped select Myron Cherry. When later asked why she chose Cherry, she responded that her choice was based on three things: he was bright, he was dedicated, and he was mean.[32]

Sinclair's 1968–1970 efforts in raising safety concerns about nuclear power were truly pioneering. In 1971 and 1972 the first large environmental coalition, the Consolidated National Intervenors, assembled around the AEC rule-making hearings on emergency core cooling systems, and it was not until 1973 that big names in public interest circles, such as Ralph Nader and the national Sierra Club, joined the opposition to nuclear power.[33]

Although much of the press coverage from Saginaw Valley newspapers was antagonistic in its descriptions of Sinclair not all of it was. The *Bay City Times*, for example, carried a sympathetic full-page story about her activities in June 1971.[34] The article, however, did run four pictures of her in her home with her children and only two pictures of her working on information to oppose the Midland Plant. One of the pictures featured Sinclair with her eleven-year-old daughter, Ann, mixing up something on the stove. In an interview for this 1971 article, she said one of the things she encountered was the occasional patronizing comment from male committee members who seemed to think it would be impossible for a woman to think these things through. She said, "They seemed to feel a fatherly obligation to set me straight." However, Sinclair never did get "set straight." Unlike the company executives, lawyers, and others who came and went during the time the Midland Plant was under construction, she was there at the beginning of the discussions of the plant in 1968 and continued to be active in the opposition to the Midland Plant and other nuclear plants through the end of the 1990s.

Dr. Connie Moll, a woman professional, local veterinarian, and business owner who served with Sinclair on the Dow Advisory Board in the 1970s, said, "Mary Sinclair was passionate, knowledgeable, unyielding and on occasion a pain in the ass." Dr. Moll went on to say that in the 1970s it was still a man's world and professional women had to be strong.[35]

In the late 1960s and the early 1970s news items about the Midland Plant reported Sinclair's comments on her concerns about the safety of nuclear power. As the 1970s progressed and through the early 1980s, a number of serious quality control and construction management problems were reported in the press, and there were continuing new estimated completion dates that were always further in the future and massive cost overruns. These problems provided Sinclair with a continuing supply of information to use in her opposition. However, independent of her work, these quality control, construction management, and cost issues were significant individual issues that had lives of their own.

In 1994, Dr. Sinclair earned her doctorate degree in environmental communications from the University of Michigan at the age of seventy-five. In 1999, she published an article in the *Toledo Journal of Great Lakes' Law, Science and Policy*, in which she wrote, "A large body of technical literature exists that documents the adverse effects of radiation on living things. The range of consequences includes cancer, reproductive failure, irreversible genetic defects, birth abnormalities and death."[36] Mrs. William Sinclair could no longer be dismissed as the typical middle-aged housewife from 5711 Summerset Drive. And to this day if you ask people who lived in Midland at the time what they remember about the attempt to build the Midland Plant, the immediate answer is Mary Sinclair.

CHAPTER 10

How to React to Midland Plant Critics

Shame, Organize, Label

What did Midlanders do when somebody threatened their town and their jobs by introducing "irresponsible controversy?" First, they shunned the person, then, with the support of the Midland Chamber of Commerce, the pronuclear groups coalesced into the Midland Nuclear Power Committee (MNPC) on October 15, 1970, in order to develop a program to gain a construction permit for the Midland Plant.[1] At their first meeting, Harold Bosscher, general manager of the Midland Division of Dow Chemical Company, explained the need for the Midland Plant, "Without it many of the [Dow Chemical's] production operations will eventually be lost to other areas in the United States if the company is to remain competitive."[2]

Bosscher's comments were a continuation of a theme Dow Chemical executives frequently expressed as the construction of the Midland Plant dragged on: the importance of the plant to Midland jobs. In March 1968, Macauley Whiting, a member of the Dow Chemical board and director of basic resources for the company, had said, "Energy-intensive chemical processes will go wherever low-cost energy exists in combination with other attributes, like raw material availability and economic transportation to market."[3] In 1977, Paul Oreffice, Dow USA president, said, "Uncertainty over the nuclear plant's future threatens jobs at Dow's Michigan

FIGURE 8. Circa 1970, the Rev. Dr. Wayne North, chairman of the Midland Nuclear Power Committee and senior pastor of First United Methodist Church, 1965 to 1973. (Courtesy of Marilyn Clark, Midland First United Methodist Church)

Division and could result in the shutdown of some units. The Division already has real problems competing with other (Dow) locations."[4]

▪ ▪ ▪

With seemingly little background in nuclear power matters, the Rev. Dr. Wayne North, senior pastor at the First United Methodist Church, was appointed chairman of the MNPC steering committee. Before coming to Midland in 1965, he had been ordained a deacon in the Methodist Church in 1943 and served in a number of positions. Usually referred to as simply Dr. North, he had received an honorary Doctor of Divinity Degree from Adrian College in Adrian, Michigan, in 1966. He had previously graduated from Dickinson College with a Bachelor of Philosophy degree and from the Westminster Theological Seminary.[5] Then in the summer of 1969 he attended Science for Clergymen, a two-week conference conducted by the Oak Ridge Associated Universities (ORAU).[6]

ORAU was a private, non-profit corporation sponsored by forty-one southern colleges and universities that conducted programs of education, information, research, and human resources development under contract with the Atomic Energy Commission (AEC). In the fiscal year ending June 30, 1971, ORAU received $4.25 million in funding from the AEC.[7] The corporation's activities allowed the AEC to advance its goal of promoting nuclear power as mandated by the Atomic Energy Act of 1954, the same act that gave the AEC its conflicting role of enforcing regulations for nuclear power.

Dr. North was adamant about the value of the Midland Plant. He said, "I'm convinced this [nuclear plant] is for the good of the ecology," and labeled some of the opponents "rabid" in their opposition. "We have 10 fossilfired [*sic*] plants, and this would take 60 tons of fly ash out of the air every day," he declared.[8] Dr. North described his views on the Midland Plant to editorial writer Bud Farrar : "There are moral and ethical implications to this issue. We have to have a nation where we can provide energy for our children. We have a moral obligation to provide employment for the people of Midland." He went on to label many people who opposed this plant misinformed on safety and radiation issues. "It's hard for people to make a transition from the emotionalism and fear generated by the Hiroshima bomb to the fact that nuclear power generation is safe and harmless."[9] As to the safety of the Midland Plant, Dr. North told Farrar, "The chances for a major nuclear accident are in the order of a billion to one."[10] In reality, the U.S. Nuclear Regulatory Commission's Rasmussen Report (WASH-1400)—completed in August 1974 and published in October 1975—said the chances of one fatality per year with one hundred nuclear reactors in operation was one in five billion.[11]

In his role with the MNPC, Dr. North reached out to ORAU for help. On October 15, 1970, he called William Pollard, executive director of ORAU. This call resulted in an invitation from Courtland Randall, the organization's chairman of information and exhibits, for a group from Midland to visit the Oak Ridge National Laboratories in Tennessee. In response, forty eight members and guests of the MNPC flew by charter jet from Midland to Oak Ridge on November 9 for a day of briefings on nuclear energy.[12] The Midland contingent's day at Oak Ridge ended with dinner and an address by Pollard who, after telling the group that he had deep concern for the environment, said that environmentalists were working against their own best interests. "I'm convinced that, in the end, they're [the environmentalists] going to realize from bitter experience that mankind can satisfy his needs for energy only thorough nuclear power. That this is the one option we have that will do minimal affront to the biosphere."[13]

One member of the Midland contingent, Edward Kern, said, "One important thing to me was learning the background these people put into nuclear power and the safety concerning it. I was particularly impressed with the environmental study and the way limits and standards for nuclear plants have been arrived at and especially the conservative nature of these limits, which have been set by groups containing mainly physicians and pathologists."[14] Based on Kern's statement, it seems that the Midlanders found that the information they received in Oak Ridge assured them that their local nuclear facility would be safe.

In addition, Dr. North's October 15 call led to a series of ORAU programs in Midland. He and Randall developed a plan for workshops that called for the MNPC—without general press fanfare—to invite small groups of fifteen to twenty to attend. The MNPC agreed to reimburse expenses not to exceed $4,000 ($26,800 in 2020 dollars). Dr. North and Randall agreed that the workshops' purpose should be to educate attendees about nuclear radiation and the technology of nuclear reactors, and not to appear to advocate the siting of nuclear reactors in Midland. Two days after the deal for the workshops was struck, two ORAU staffers left for Midland in an "unmarked ORAU station wagon." They were later joined in Midland by two other staff members.[15]

On November 28, the *Midland Daily News* reported that "the 'students' of each workshop were invited by the MNPC, which kept class sizes small in order to promote the maximum participation of each individual. An attempt to reach a broad cross section of citizens was made as housewives, ministers, tradesmen, professional persons and those in service type jobs comprised a typical class." The article went on to say that the MNPC, in its efforts to support the construction of the Midland Plant, felt "its first commitment has been to learn the basic facts of nuclear power and how it fits into our lives today in terms of needs, safety, and environmental effects."[16]

Upon returning home to Oak Ridge, the team that had given the Midland workshops expressed a rather dim view of their efforts. In general ORAU staff members found that workshop participants had very little comprehension of the nuclear establishment's technical vocabulary: "They found Midland citizens in some ways unusually sophisticated, but their grade level in terms of understanding such terms as neutron, fusion and radiation was poor ninth grade general science. Those who attended the workshops were mostly middle aged or older, and they represented a lost generation in scholastic background on nuclear energy."[17]

The MNPC efforts were not restricted to these workshops. Bumper stickers were printed, space was bought for ads in the local newspaper, and members of the MNPC released statements to the press to rebut Mary Sinclair.[18] After she spoke to the Chamber of Commerce in Freeland, a small community twelve miles from Midland, in August 1971, Mrs. Sidney N. Smock and Ned S. Arbury of the MNPC said,

> Mrs. Sinclair, in commenting on the shutting down of the propylene oxide plant by the Dow Chemical Company, stated nobody is chasing Dow out. We refer Mrs. Sinclair and the Saginaw Valley Nuclear Study Group to Dow's news release of

> August 16 wherein Dow spelled out the much higher (twice) cost of raw materials and energy, in Midland with reference to propylene oxide production, over similar costs in Sarnia, Canada and Freeport, Texas. Dow also noted that in addition to current high costs for energy, the uncertainty about when lower cost energy might be available was a major factor in the shutdowns.[19]

This was not a very veiled implication that Sinclair was indeed chasing Dow Chemical out of Midland.

One thing did not happen, though, with the MNPC's efforts to promote nuclear power and rebut Sinclair's efforts. Dr. North refused to debate Sinclair, explaining, "If you don't know any better, what she says sounds very logical. Only more confusion would come from debate."[20]

■ ■ ■

The MNPC's concern in 1970 and 1971 that the Consumers Power Company had not received a construction permit for the Midland Plant was well founded. The utility had said that it expected a construction permit by the beginning of 1971, but the hearings for it did not even start until December 1970. In July of that year Consumers Power had obtained a variance from the AEC to allow limited construction activity on the site of the Midland Plant, but in November, reporting that it had completed most of the construction that it planned to do before receiving a construction permit, the company stopped all work on the site. With this announcement, Consumers Power said it would not resume construction until it obtained its construction permit from the AEC.[21]

What was behind these delays? By the summer of 1970, it had become apparent to Consumers Power that intervenors in AEC hearings could delay approvals. In fact, a young lawyer named Myron Cherry was causing a continuing delay in Consumers Power's receiving its last necessary approval (an operating license) from the AEC to run the Palisades Plant.[22] Consumers Power had not foreseen the problem of intervenors. There had been no opposition to the granting of the construction permit for the Palisades Plant, and the hearings had been perfunctory. From the time they announced the Palisades Plant until the construction permit was granted only took fifteen months: January 1966, they made the announcement; June 1966, they filed an application with the AEC for construction permit; in February 1967, there was no opposition filed to oppose the permit; and in March 1967, the AEC approved the construction permit.[23]

But in 1970 Cherry delayed the AEC proceedings by challenging the issuance of an operating license for the Palisades Plant. The AEC licensing involved a two-step process: first, a construction permit was issued, and then an operating license needed to be issued before the reactor can start generating electricity. Although the AEC had issued a notice in March of 1970 that it intended to give Consumers Power this license, Cherry managed to drag the hearings out until 1971 with his never-ending questions.[24] Consumers Power finally agreed in March 1971 to spend an additional $29 million and incur added costs of $5 million per year for operating expenses. The utility defended the additional cost with the comment that management considered these additions mandatory in order to permit timely operation of the plant.[25]

By this time, the company understood that it was going to have to factor intervenors into its plans. In its 1971 annual report, the company stated that construction on the Midland Plant was halted in November 1970 after it became apparent that intervention by opponents would greatly prolong the process of obtaining a construction permit.[26]

In light of these events, MNPC realized it needed to undertake more ambitious programs. Midland needed certainty that inexpensive energy would soon become available in town so that Dow would not start moving their chemical manufacturing away, and those bureaucrats in Washington obviously needed a good dose of Midwestern boosterism to create political pressure to get the AEC construction license issued for the Midland Plant. The Speak Up Now rally discussed earlier was the MNPC's answer. Sponsored by the MNPC and financed with $32,000 ($200,000 in 2020 dollars) collected by the Midland Chamber of Commerce, Speak Up Now was a major display of public concern that if the nuclear facility were not built, the town's economy would suffer.[27]

Efforts did not end with the Speak Up Now rally. Afterwards, a large delegation from Midland delivered the ten-by-twenty-foot "License for a Nuclear Power Plant" to Washington embellished with fifteen thousand signatures. Ned Arbury listed the members of the delegation sent to deliver the billboard-sized license in an October 22, 1971, memorandum that detailed the hotel reservations.

a. Three twin-bedded doubles and one double bedroom suite: James A. Kendell, Robert L. Parker, Alan W. Ott, Bruce R. Benway, Richard T. Miller, Julius Blasy, Robert B. Chatterton, and Ned S. Arbury. Note: (If the loud snorers and sleep talkers will identify themselves in advance, we will house accordingly.)

b. Three single rooms: Mrs. Sidney Smock, Dr. Wayne E. North, (Chairman of the delegation), and Simon Chapple (Ass't to President of AFL-CIO) with William Marshall occupying the room beginning on 10/26.[28]

In addition to these people, the Midland delegation was accompanied by Macauley Whiting, director of Hydrocarbons and director of Dow Chemical; James F. Maddox, manager of the Nuclear Energy Project at the Dow Chemical; Russell Youngdahl, senior vice president, Consumers Power; and William Kessler, nuclear project manager of Consumers Power. They held a whirlwind of meetings in Washington, D.C., over a period of three days, October 26 to 28, 1971. The group met with staff representatives of the AEC, representatives of the Environmental Protection Agency, members of the Michigan Congressional Delegation, and members of the Joint Committee on Atomic Energy. All of these meetings were an attempt to speed up the issuance of a construction permit for the Midland Plant.

The front-page article in the *Midland Daily News* on October 29 said the paper had learned in an exclusive report that in an October 27 meeting with officials of the Environmental Protection Agency and the Council on Environmental Quality, the Midland Plant had been placed in a "top priority" status. Dr. North said, "There is no doubt our visits with officials of the AEC, Presidential advisors, the joint Commission on Atomic Energy, and Michigan's elected officials have enhanced our prospects for an early decision on the [construction permit] application."[29]

However, this "top priority" that the delegation thought they had achieved for the Midland Plant's construction permit did not seem to make much difference. The day after Dr. North gave his optimistic report, the AEC said perhaps it could finish reviewing the Midland Plant's emergency core cooling system (ECCS) by December 1971 at the earliest and that reevaluation of the ECCS material was one part of the administrative delays currently holding up the hearings (the other being environmental reviews). In addition, the AEC said that the licensing process could only move forward if regulatory staff were available when needed to review the ECCS submission.[30] At the end of October 1971, it was apparent that the MNPC's Speak Up Now rally and its trip to Washington, D.C., were not going to make a difference.

Another problem—with which the MNPC chose not to engage—was the public hearing for the construction permit, which the opponents had hired lawyers to contest. The hearings pitted the attorneys for the opponents of the Midland Plant against several Consumers Power attorneys, including Robert Lowenstein from the firm of Lowenstein and Newman, and Dow Chemical attorney Milton Wessel.

Although they filed frequent complaints with the AEC regarding the prolongation of the permit process, Lowenstein and Wessel were ineffective in shortening the duration of the hearings.[31] While it took only fifteen months to get a construction permit for the Palisades Plant, it took the Midland Plant five years from the date of the announcement in December 1967 to receive its construction permit.

In addition to stopping construction of the Midland Plant in 1970 while waiting on a construction permit, Consumers Power also decided to stop all engineering and design work. Therefore they did not keep up with new design requirements that were issued by the AEC during the time construction stopped until it was resumed in February 1973. When it restarted, it had to play catch up on the engineering and design work for plant safeguards and control systems. The utility found that whereas the engineering and design work had been estimated to be 49 percent complete in December 1970, with the new requirements, by August 1973 that estimate had plummeted to only 25 percent.[32] The cost of the Midland Plant, initially projected to be $267 million, had risen to $765 million by the time the construction permit was finally granted in December 1972.[33]

In spite of the lack of success in speeding up the construction permit for the Midland Plant in 1970–1971 and indications that costs were going to increase and that construction would fall behind schedule, Midlanders continued to support the building of the nuclear facility after the early 1970s.

CHAPTER 11

Midland Pride

Why Midlanders Supported the Midland Nuclear Plant

Context is needed in order to understand the October 1971 Speak Up Now rally, attended by 40 percent of the town's population. Independent of whatever reasons Dr. Wayne North, the Midland Nuclear Power Committee (MNPC), and participants at the rally put forth, there were fundamental reasons for Midlanders to support the nuclear facility. Local people's view was that the Dow Chemical Company helped win World Wars I and II and would provide what the country needed in peace and in war. The company was also led by respectable people who lived in Midland, whom they saw at church and in community activities; provided meaningful employment; continued to develop important products that would allow the company to grow; and helped make the town a nice place to live.

When Dow Chemical said it needed the Midland Plant, Midlanders responded positively because they took pride in the company and in their town. It was a good place to live even if there was a little coal soot, and they did not want to lose their jobs and have to move. By the 1960s Midland had transformed itself from the dying lumber town without any outstanding identity that it had been when Dow Chemical arrived. While nearby Bay City and Saginaw still had their aging, dark brick and stone churches built in the 1800s, Midland had fifteen new mid-century modern churches that had been designed and constructed between 1941 and 1968. Bay City and Saginaw still had their aging Victorian mansions, but Midland had over three hundred mid-century modern residential structures that had been built between 1933 and 1967.[1] One needed only to drive around Midland in the 1960s to be proud that one lived in in a modern, up-to-date city.

The amiable connection between the people of Midland and Dow Chemical—and indeed the Dow families—was a long-term, close relationship. It is often said that many companies in America come to town, make money, leave town, and take their money with them. The founders and the executives guiding Dow Chemical up to and through the period of this story did not. When World War I caused the company to drastically increase its workforce to supply chemicals for the war effort, Herbert Henry Dow became concerned that the newly expanded workforce did not have the facilities for recreation. To meet this need the Midland Community Center was founded in 1916 with a fund of $83,735, of which Dow contributed $37,010 out of his own pocket, other members of the Dow family contributed $1,760, Dow Chemical contributed $22,000, and seventy-three other persons contributed $22,965.[2] Also, Dow brought a Japanese gardener to town and made him available for free to anyone who wanted to improve their yard or garden.[3]

In 1936 Willard Dow, Dow's first son and at the time chief executive of the company, started the Dow Male Chorus with a donation to rent a piano and pay a director. Later he bought an old Nazarene Church building near the company headquarters and remodeled it as the Dow Music Building. The company's music program grew to include a symphony orchestra, a girls' chorus, a boys' choir, a madrigal group, and a choral society. In 1944, the Dow Music Department staged its first big show, a performance of the Gilbert and Sullivan operetta *H.M.S. Pinafore*. Shortly after that production it expanded to include four professional musicians, and more than five hundred Dow Chemical employee volunteers and their family members participated in its programs. In the spring of 1961, Dow Chemical got out of the music business, but a Midland Music Foundation was launched and was able to lease the Dow Music Building for a token payment.[4]

In 1950, Grace A. Dow, chairman of the Midland Library Board for thirty-two years, announced that the Herbert H. and Grace A. Dow Foundation (which she established in 1936 in memory of her husband) would build a new public library in town on the property that had been her family's apple orchard.[5] The Midland Center for the Arts (MCFTA) held its first performance with a Christmas concert on December 5, 1970. This building, designed by architect Alden B. Dow, another son of Dow, includes a 1,538 seat auditorium with a seventy-foot stage and orchestra

TABLE 1. Herbert H. and Grace A. Dow Foundation grants to Midland beneficiaries in 1975.

BENEFICIARY	AMOUNT (1975 DOLLARS)
City of Midland Ice Rink	$1,529,945
Midland Community Center	$849,763
Dow Gardens	$467,303
Chippewa Nature Center	$287,237
Midland Beautification Trust Greenbelt	$254,601
Midland Center for the Arts	$165,764
Midland Community Foundation	$13,000
United Community Fund of Midland	$10,500
Total	*$3,578,113*

Herbert H. and Grace A. Dow Foundation, *1975 Annual Report* (Midland: Herbert H. and Grace A. Dow Foundation, Spring 1976).

lift, a 426-seat theater, a ninety-four–seat lecture hall, a Hall of Ideas (now named the Alden B. Dow Museum of Science and Art) with changing exhibits, historical exhibit areas, and three art galleries. There are studios for working potters, painters, weavers, photographers, and metalsmiths; dining and lounge facilities; rehearsal halls; meeting rooms; and offices. Although three thousand individuals, businesses, and industries contributed $1.5 million to the construction of the MCFTA, it was also made possible with a $4 million gift from the Herbert H. and Grace A. Dow Foundation and $2.5 million given by Alden B. and Vada Dow. In addition, the Herbert H. and Grace A. Dow Foundation contributed $2.5 million for an endowment for the Center for the Arts.[6] And the Dow Botanical Gardens, spanning 110 acres in Midland, also exists because of the Herbert H. and Grace A. Dow Foundation. Table 1 lists the foundation's grants for the Midland community in just the year 1975.

During the late 1960s, Dow Chemical leadership felt that the public face of the plant presented a misleading image of the work that was carried on inside. By 1972, a plan was in place to beautify the areas between the Dow Plant and Poseyville Road, where the public had a view of the Dow Plant. Helen Dow Whiting, granddaughter of Herbert Henry Dow, along with her husband and Dow Chemical employee Macauley Whiting, generously gave the necessary funds for the park from their private foundation, and it was named after them, Whiting Overlook Park.

Carl Gerstacker, a long-term employee who rose through the ranks to be the chairman of Dow Chemical's board from 1960 until 1976, explained the Dow family's community philanthropy:

> Midland was a terrible place to live when Herbert Henry Dow came to town. It was a leftover lumbering town, dying on its feet, and Main Street was mainly drinking places. I finally got through my thick skull that the reason he was doing all this was that he wanted to bring the best people he could find to Midland, the best chemists, the best people for his company, and the only way he could get them to come here and stay here was if this was a good place to live.[7]

It may have been enlightened self-interest on the part of the management of Dow Chemical that made Midland a nice place to live, but the people of Midland appreciated these efforts on their behalf.

Local leaders returned Dow Chemical's commitment with their own. Dr. North's activities to support the Midland Plant are not a unique example of the town's clergy supporting Dow Chemical's position. Take, for example, the issue of the manufacturing of napalm. The Midland Clergy Association had written a letter and read it at the Dow Annual Meeting back in 1968 supporting Dow Chemical's manufacture of napalm for use in the Vietnam War. Signing that statement were clergy from the First Baptist Church, Gordon Street Church of God, First Methodist Church, Holy Family Episcopal Church, and Memorial Presbyterian Church.[8] This letter was a response to 260 demonstrators organized by an outside group called The Clergy and Laymen Concerned about Vietnam, which protested against napalm production at Dow Chemical's 1968 annual meeting in Midland. After making clear that the Midland ministers agreed with the protesters that "war is hell," the statement said:

> But we also affirm the thesis that, if the management believes that under the present situation the prosecution of the war must continue until negotiations effect a peaceful solution, which is deemed to be for the welfare of Southeast Asian people and the future security and peace of the world, it would be recreant in its responsibility to our government not to produce and make available an instrument of warfare, however horrible (for all instruments of war are horrible) its effects may be, even in its most discriminating use.[9]

Napalm was not a new weapon at the time of the Vietnam War. It was invented on Valentine's Day 1942 at a secret Harvard war research laboratory by organic chemist Dr. Louis Fieser. In 1945 napalm was used to bomb sixty-six of Japan's largest cities.[10] U.S. general Curtis LeMay, who directed the attack on Tokyo in 1945, said, "We scorched and boiled and baked to death more people in Tokyo on the that night of March 9–10 than went up in vapor at Hiroshima and Nagasaki combined."[11] In 1968 it was reported that the U.S. Air Force had dropped more than 100,000 tons of napalm on Vietnam since 1963.[12] The use of napalm inspired one of the anti-war chants of 1968: "Hey! Hey! LBJ! How many kids did you burn today?"[13] In 1972 a picture of a nine-year-old naked Vietnamese girl burned with napalm would become the iconic picture of the U.S. involvement in the Vietnam War.

But the Midland clergy wrote, "We would strongly resist the implication, if any, by the members of The Clergy and Laymen Concerned about Vietnam that the Dow directors are, or have been, lacking moral integrity or have been apostate in their responsibility to humanity in their decision to produce and supply an element of warfare in the hour of our country's involvement in war."[14] Dow Chemical's response to the protesters was

> We intend to continue making it [napalm] because we feel that so long as the United States is sending men to war it is unthinkable we would not supply the material they need. We as a company have made a moral judgement on the goals of our government, and we support them. We may not agree as individuals with every decision of every military or governmental leader—but we regard these leaders as men trying honestly and relentlessly to find the best possible solutions to very complex international problems.[15]

In 1968 Dow Chemical and the people of Midland came down strongly in support of the U.S. government and providing what it needed for the war. Fifty years later, Dow Chemical's manufacture of napalm was still being cited in the debate about whether large technology companies should provide their expertise and products to the U.S. military. In a March 4, 2019, op-ed in the *New York Times*, Kevin Roose noted that employees at Microsoft and Affectiva, an artificial emotional intelligence company startup, did not want to work with the military because their core values of integrity and respecting people's privacy were more important than the need of the military to stay current with its adversaries. Roose wrote that in 2019 companies should avoid the negative treatment that Dow Chemical had received for producing napalm because it could cause the share price of the company to drop.[16] Dow Chemical's share price and personal privacy ideas did not inhibit the company in 1968. In Midland Dow Chemical management said that if the U.S. government needed napalm for the Vietnam War, the company would supply it. The town that Dow Chemical had built was a great place to live, and Midlanders believed in the moral integrity of the leaders of the Dow Chemical Company—if the government needed napalm, Dow Chemical should supply it.

■ ■ ■

Things were not great everywhere in Michigan in the 1960s. Midlanders looking south in the summer of 1967 to the auto manufacturing towns of Flint and Detroit saw civil disruptions. Detroit's population in 1967 was 38 percent Black, and for decades a nearly all-White police force had brutalized and humiliated people of color, while the political structure had marginalized the voices of Black residents. On July 23, 1967, a 3:30 a.m. police raid on an unlicensed, after-hours bar known as a blind pig on Detroit's west side led to a riot. Forty-three people died, 1,189 were injured, and 7,231 were arrested. Stores looted or burned included 611 food markets, 537 cleaners, 285 liquor stores, and 240 drug stores.[17]

In 1967 there were no civil disruptions in Midland. No 1967 record of bad police behavior in Midland existed. Midland had good schools and good housing available for all. All of these characteristics are not a testimony of its being a place free of racial inequity, though, as Midland had virtually no Black population. The great migration of poor Blacks from the rural southern United States to the auto manufacturing centers in southeastern Michigan that started about 1916 had not reached as far north as Midland. It was not until 1960 that Linneaus Dorman became the first Black professional Dow Chemical moved to its Midland headquarters, and for four years the Dormans were the only Black family in town. So unaccustomed were local hospital staff to registering babies in any racial category but White, that Linneaus and Phae Dorman's son's birth certificate had to be corrected because both baby and parents were misidentified as White.[18] In the 1960s as Dow Chemical began hiring more Black professionals, Dorothy Fauver hosted forums to engage the newcomers and educate those unaccustomed to living and working with people of color.[19] Not everyone appreciated Fauver's efforts. She received hate mail, both anti-Black and anti-Semitic, and a cross was burned on her lawn.[20]

The Speak Up Now rally and the other activities by the MNPC expressed support for the Midland Plant. They also supported Dow Chemical and people's appreciation for the town that the company built. But the nuclear facility's supporters' unsuccessful attempts to exert political pressure to influence the approval of the construction permit for the Midland Plant did not address the regulatory and legal challenges to the plant. As it turned out, it was the damage to the hull of a boat belonging to a senior partner in a Chicago law firm that made a critical difference in the length of the construction permit hearings and the later legal challenges to the Midland Plant.

CHAPTER 12

Myron Cherry

Bright, Dedicated, and Mean

When Jim Campbell was president of Consumers Power Company in 1970, he thought the company was on a successful learning curve for producing nuclear power. The company was operating the Big Rock Point Plant, fueling the Palisades Plant, and planning for the Midland Plant. The public hearings for the Palisades Plant facility, however, allowed young lawyer Myron Cherry to take advantage of a completely different learning curve—one that provided him education and training in bedeviling the builder of a nuclear plant.

Originally, a senior partner in his Chicago law firm dispatched Cherry to look at Consumers Power's Palisades Plant in 1969. The partner was upset with the damage his pleasure boat had sustained when it struck debris from an old PT boat that Consumers Power had sunk to form a breakwater in Lake Michigan during the construction of the Palisades Plant. The utility refused to pay anything to repair the partner's boat and told the partner and Cherry that anything related to the material in Lake Michigan was highly classified information.[1] The partner then told Cherry, "Go over to Michigan for few days and give those sons of bitches a hard time."[2] Cherry recalled, "Had Consumers Power agreed to repair the hull of that partner's boat (probably for less than $15,000), my career in opposing nuclear power never would have happened."[3]

FIGURE 9. July 13, 1971, lawyer Myron Cherry, pencil at the ready, attempting to get the construction permit for the Midland Plant denied, challenges a witness. (Courtesy of the *Midland Daily News*)

Cherry did find a way to "give the sons of bitches a hard time." He got himself hired as the counsel for local environmentalists and fishermen who feared that the Palisades Plant would damage marine life in the area, and he ended up spending nine months representing them in the operating license hearing conducted by the Atomic Safety and Licensing Board (ASLB) for the Atomic Energy Commission (AEC). When Cherry posed a long list of environmental questions about the Palisades Plant, regulators said they did not have authority to rule on environmental matters. He then shifted to raising hundreds of safety concerns. He later said, "We were determined to drag that hearing out until something was done for my clients."[4] Consumers Power finally agreed in March 1971 to install $29 million worth of environmental safeguards at the plant, explaining the decision by saying that with the plant standing idle they needed to do something.[5] In 1978 Cherry told the *Wall Street Journal* that in building a nuclear power plant, a utility should have an experience "as painful and expensive as possible."[6]

Bright, dedicated, and mean is how Mary Sinclair described Cherry.[7] She decided to hire him to represent the Saginaw Valley Nuclear Study Group in the Midland Plant construction permit hearings, which began on December 1, 1970.[8] As mentioned earlier, the group's intervention of the Saginaw Valley Nuclear Study Group quickly expanded to include five more groups, forming the Saginaw Intervenors.

A separate group called the Mapleton Intervenors was also active in opposing the granting of the construction permit for the Midland Plant—this group consisted of eight homeowners and businessmen from the tiny community of Mapleton just downstream on the Tittabawassee River from the proposed nuclear facility. This group was interested primarily because of the nuisance the plant might cause and its effect on property values. Sinclair declined to consolidate with the Mapleton Intervenors because she saw her group as being national in character with a broader range of issues. Cherry, however, did share documents with the Mapleton Group during the construction permit hearings.[9] Its lawyer, Irving Like of New York, shared Cherry's belief that, in spite of the fact that the AEC had a bias for industry, the intervenor could use the agency procedure for "education of the public and consequent public support."[10] Like summarized his views clearly in the title of his article in *Ecology Law Review*: "Multi-Media Confrontation—The Environmentalists' Strategy for a No-Win Agency Proceeding."[11]

Prior to the start of the Midland Plant construction permit hearings on December 1, 1970, Cherry and supporters of his intervention held a press conference at Midland's Holiday Inn. He declared that the days were gone when the public was not involved in decisions such as those involving a nuclear plant's construction. Indicating that his group was confident about succeeding in bringing the issues before the public, Cherry said, "Time and momentum are on our side."[12] Although influential national groups had not yet declared their opposition to nuclear power in 1970, the Saginaw group's attorney was apparently making reference to a large number of ad hoc local groups being organized all over the country in the 1960s and 1970s to oppose nuclear power plants close to where their members lived.[13]

Public hearings for a construction permit for a specific site are much like a trial in which the permit applicant bears the "burden of proof." Although one might think that a permit hearing would be restricted to technical and scientific details of the specific reactor proposed for installation, many other topics are reviewed. Because of its trial-like nature, lawyers can create much ado about a whole host of issues: site security, evacuation plans, safe levels of radiation beyond the site, procedural issues, public notifications, and waste handling.

Gordon Heins, one Consumers Power witness in the construction permit hearings, said he remembers that Cherry was an unpleasant person who didn't seem to know much.[14] But Cherry did not see a lack of background knowledge as an impediment to his efforts. When an interviewer asked about his participation in hearings, Cherry said, "There are some things I'm not so good at, but I'm absolutely fantastic at asking questions."[15] Given the enormous legal playground that the approval of a nuclear plant created, he had plenty of opportunity to demonstrate his question asking skill.

Cherry was also a master in setting the stage for the Midland Plant construction permit hearings. Prior to one of the hearings, he dispatched Peter Sinclair, Sinclair's eighteen-year-old son, to get into the cafeteria of Midland's Jefferson Intermediate School, where the hearing would be held, and rearrange the seating. He put Cherry's place front and center and crowded the seating for the Consumers Power people into a corner.[16] Cherry also loved to heckle technical witnesses. Once when a Consumers Power witness lapsed into technical jargon, Cherry walked over to a woman in the hearing room and asked her if she understood the witness. She said she didn't. "Neither did I," the lawyer said. He then told the witness, "This is a public hearing. Say it so the public can understand you."[17]

Cherry also enlisted others to help with his performances at hearings. One of them was Dr. Robert Asperger, an inorganic chemist working for Dow Chemical Company in the 1970s. Asperger had once taught radiation chemistry in an AEC-sponsored program and had become an outspoken critic of using nuclear power to generate electricity. In addition to working with Sinclair, he wrote to the bond rating departments of Moody's Investors Service and Standard & Poor's Corporation to make the case that a utility's bonds were not a low-risk investment if the utility owned or was proposing to build a nuclear power plant.[18] In 2016 Asperger recalled that one of his jobs had been to sharpen pencils for Cherry. These sharpened pencils were not for taking notes. The lawyer used them in interrogating witnesses, moving in as close as he could and then breaking a pencil to disturb and distract the person.[19] He would also visit the table of a Consumers Power lawyer at lunch and snap a sharpened pencil for every document he thought his opponent had improperly withheld.[20]

At the construction permit hearings held at Midland's Jefferson Intermediate School in July 1971, Cherry, pencil at the ready, challenged witnesses on the adequacy of emergency and evacuation plans for the proposed nuclear plant. He began by stating that the evacuation plan Consumers Power had filed for Midland was just a photocopy of the one they had submitted for the Palisades Plant. Then, in a cross-examination of three witnesses that lasted nearly seven hours, he probed into the creation of the total emergency plan created for the Midland Plant. He also inquired about the effects of weather conditions on chemical spills like a recent bromine fog on Midland's Saginaw Road and the impact of the interconnection of some plants on evacuation procedures.[21]

Cherry waged guerilla warfare in the hearing room. For example, having discovered that one of the Consumers Power lawyers could not tolerate loud noises, Cherry would shout whenever the two were called to the bench. That lawyer eventually withdrew from the courtroom.[22] At one point in a proceeding, Cherry suggested that an attorney for the Nuclear Regulatory Commission (NRC) "should be sued by the public for his last two years' pay since he hasn't done anything to earn it."[23] In yet another case, Cherry commented to a Consumers Power public relations executive, "You are so ugly your wife deserves combat pay."[24] In addition to these theatrics, the Saginaw Intervenors' attorney submitted over three hundred interrogatories (written questions on such issues as energy conservation, emergency evacuation plans, and radiation safety) that the parties in favor of the Midland Plant had to take the time to address. Eventually there were nine thousand pages of testimony in the construction permit hearings.[25]

■ ■ ■

It should be noted, however, that the extended time the construction permit hearings took was not all due to Cherry. In 1970, President Richard Nixon signed the National Environmental Protection Act (NEPA). In 1971 in *Calvert Cliffs' Coordinating Committee, Inc. v. United States Atomic Energy Commission* (*Calvert Cliffs'*), the courts interpreted the NEPA as having created a cause of action against federal agencies that did not comply with the act's directives. Essentially, this case made federal agencies accountable for their impact on the environment. Specifically, the courts ruled that the AEC's NEPA compliance rules were insufficient to fulfill the act's environmental impact assessment duties. For the Midland Plant *Calvert Cliffs'* meant that Consumers Power had to prepare an environmental impact statement.[26] The resulting detailed environmental study, with amendments, ran to more than 1,500 pages and took ten months to complete.[27]

Another reason for the prolonged hearings was that the AEC had issued new guidelines for emergency core cooling systems (ECCS) in 1971 because tests it had conducted in 1970 and early 1971 suggested that during a loss-of-coolant-to-nuclear-reactor incident, adequate cooling water would not be available, which in turn could lead to a reactor meltdown.[28] The new design criteria required a reevaluation of the Midland Plant's system, a job that took five months. In addition, concerns about the new criteria caused the AEC to hold hearings on ECCS in 1972, which halted Midland's construction permit hearing for ten days so the AEC's Atomic Safety and Licensing Board personnel could attend the ECCS hearings.[29]

Consumers Power frequently blamed Sinclair and the other opponents of the Midland Plant for delays in construction, but it did acknowledge that its own submittals to comply with the new environmental and ECCS requirements contributed to slowing things down as well. AEC staff workloads also prolonged the review process. Consumers Power said in December 1972, "With AEC staff personnel engaged with reviewing ECCS designs and environmental reports for a host of plants around the country, it wasn't until early in 1972 that the AEC's reviews of Consumers Power's submittals were completed."[30]

In the end, the construction permit was eventually granted on December 15, 1972. In Midland, many felt that the construction permit hearing had provided adequate examination of the safety of the facility and that construction would proceed to complete it. Alan Ott, president of Midland's Chemical Bank, said, "Nobody in this community wants an unsafe plant and I'd have to say that no one has ever proposed that. Mary's items about safety—I think she has provided a service—but you operate in a system and you have license procedures, and you have supposedly all the safety checks to build a safe plant [so] you have to assume all those things are correct."[31]

But the fact that Consumers Power was finally able to receive a construction permit did not end the legal and regulatory challenges. Having checked the box of administrative appeal (the precise requirement is "after all possible administrative remedies have been exhausted"), Cherry was then free to appeal in the courts against the granting of the construction permit. He filed a petition with the United States Court of Appeals, District of Columbia Circuit (D.C. Circuit) in July and August 1973. The names of the petitioners on the case were Nelson Aeschliman (the Mapleton group) and Saginaw Valley Nuclear Study Group et al.

The Atomic Energy Act of 1954 and the Administrative Procedure Act of 1946 made it possible to appeal the granting of the construction permit, and legislation and court opinions in the early 1970s appeared to favor Cherry's appeal. Section 101 of NEPA requires the federal government to use "all practicable means and measures" to restore and maintain environmental quality to the overall welfare and development of man. At the time the law represented a broad-based interest in protecting the environment, passing with a 372–15 vote in the House of Representatives (including 164 Republicans voting for the bill) and by a unanimous vote in the Senate.[32] NEPA was then signed into law by President Nixon (a Republican).

As noted earlier, the law soon affected the effort to build nuclear power plants because of the *Calvert Cliffs'* decision. Prior to this ruling, the AEC had continually asserted that it had no statutory authority to concern itself with the adverse environmental effects of its actions. Now, however, the AEC's hands were no longer tied as it had claimed when Cherry tried to use environmental arguments in the 1970 Palisades Plant's operating license hearings. The commission was not only permitted but compelled to take environmental values into account. Perhaps NEPA's greatest impact was to require the AEC and other agencies to *consider* environmental issues just as they consider other matters within their mandates.[33]

Cherry's choice of the D.C. Circuit to file the Saginaw Intervenors' appeal of the Midland Case was undoubtedly based on its reputation as a liberal court that frequently ruled in favor of environmental and consumer groups.[34] As a matter of public policy formation, the circuit's 1971 *Calvert Cliffs'* ruling signaled that it was willing to hold agencies like the AEC responsible when, in the court's view, they did not completely investigate the environmental impacts of their decisions. In addition to the *Calvert Cliffs'* ruling, Judge David Bazelon, the D.C. Circuit's longtime chief judge, in the 1971 *Environmental Defense Fund v. Ruckelshaus* had said, "Courts are increasingly asked to review administrative action that touches on fundamental personal interests in life, health, and liberty. These interests have always had a special claim to judicial protection, in comparison with the economic interests at stake in a ratemaking or licensing proceeding."[35] Bazelon was renowned for his commitment to using the law as a tool to make regulatory agencies do a detailed review to support their actions. In the words of one of his colleagues, "[a] restrained judge he was not." He was also famous for the lunches he organized in a backroom of Milton Kronheim's liquor warehouse, where liberals from across Washington, D.C., would gather.[36]

Cherry's appeal to the D.C. Circuit was successful. As it in the *Calvert Cliffs'* case, the D.C. Circuit found fault with the AEC's action in granting the construction permit for the Midland Plant. On July 21, 1976, the judges issued their decision and told the AEC to have the Advisory Committee on Reactor Safeguards (ACRS) clarify its report on reactor safeguards, reconsider the impact of the plant on energy conservation, reevaluate the Dow Chemical's need for the facility, and undertake appropriate consideration of waste disposal beyond the simple statement that wastes would be shipped to unidentified offsite disposal areas.[37]

But Cherry's victory in the decision of the D.C. Circuit Court in July 1976 did not stand. Consumers Power sought and was granted a review of the ruling by the U.S. Supreme Court. The Midland case was coupled with the Vermont Yankee case, and the case names that appeared on the docket were *Vermont Yankee Nuclear Power Corp. v. Natural Resources Defense Council, Inc., et al. and Consumers Power Company v. Aeschliman, et al.* Justice William Rehnquist delivered the unanimous opinion of the U.S. Supreme Court on April 3, 1978.[38] On every point that the D.C. Circuit had remanded the construction permit for Midland Plant to the AEC, the Supreme Court did not just disagree—it strongly disagreed.

On the issue of clarifying the ACRS's report on reactor safeguards, the court ruled that there was nothing to support a court's ordering the commission to take that step or requiring the ACRS to give a short explanation, understandable to a layman, of each generic safety concern.[39] As to reconsidering the impact of energy conservation, the justices ruled that since energy conservation was a novel and evolving concept, only in the future could an adequate record be developed.[40] On the evaluation of Dow Chemical's need for the plant, they found that the firm's need for steam was not an independent basis for vacating and remanding the commission's decision and that the AEC had considered the changed circumstances and refused to reopen the proceedings at least three times.[41] And finally, although the Supreme Court remanded the nuclear waste issue for further consideration, it said that NEPA did not require the NRC to undertake any procedures beyond the minimum requirement required by the Administrative Procedure Act.[42]

In concluding the opinion, the Supreme Court issued a strong rebuke to the D.C. Circuit judges:

> Consumers Power first applied in 1969 for a construction permit—not even an operating license, just a construction permit. The proposed plant underwent an incredibly extensive review. The reports filed and reviewed literally fill books. The proceeding took years, and the actual hearings themselves over two weeks. To then nullify that effort seven years later because one report refers to other problems, which problems admittedly have been discussed at length in other reports available to the public, borders on Kafkaesque. Nuclear energy may someday be a cheap, safe source of power or it may not . . . Time may prove wrong the decision to develop nuclear energy, but it is Congress or the States, within their appropriate agencies which must eventually make that judgement. In the meantime courts should perform their appointed function. NEPA does set forth significant substantive goals for the Nation, but its mandate to the agencies is essentially procedural.[43]

On the issue of nuclear waste disposal, the D.C. Circuit had made clear that the NRC's assurances on waste disposal were not based on an informed decision-making process. The problem arises because although nuclear fuel contains a lot of energy (one uranium fuel pellet the size of a fingertip creates as much energy as 1 ton of coal, 149 gallons of oil, or 17,000 cubic feet of natural gas), new fuel rods need to be placed in a reactor from time to time.[44] Since uranium fuel breaks down into several highly radioactive byproducts, these products are present in the old fuel rods. Waste disposal was not an abstract issue with the Midland Plant. Each of its two reactors was to have in its core 90 tons of fuel. About one-third of this fuel was to be replaced each year, leaving Consumers Power with the problem of where to store 60 tons of spent but still highly radioactive fuel each year.[45] With the failure of the Supreme Court to require a reasoned decision by the NRC on what to do about nuclear waste in the Midland case, the issue of the permanent storage of nuclear waste from power production was left to fester. As of 2020 a total of about 80,000 metric tons of nuclear waste is stored at power plants in thirty-three states.[46]

▪ ▪ ▪

Were there any benefits to Cherry's court appeals? It might seem there were not because in addition to his failure to convince the Supreme Court to cancel the Midland Plant's construction permit, he was also not even able to get construction stopped during the time his appeal was under consideration by the courts. To see the value of Cherry's appeals, it is essential to focus on the fact that the Supreme Court did not act until April 1978, more than five years after the construction permit had been granted. People in Midland wanted to believe that construction of the Midland Plant had been assured when the permit was approved in 1972, but the opponents continued to point to the ongoing court cases challenging it. In the absence of legal certainty for the Midland Plant until 1978, Sinclair continued to publicize the problems with nuclear waste, plant safety, the cost, and the need for new electrical generation as unresolved issues.[47] Then in 1979 when the movie *The China Syndrome* and an accident at the Three Mile Island Nuclear Plant made nuclear power a national issue, the arguments made during the years Cherry dragged the Midland case through the courts suddenly seemed much more credible.

James G. Keppler, Midwest director for the NRC, summed up another result of Cherry's efforts. He conceded in 1978 that while he might not approve of Cherry's style, the lawyer's criticisms had pushed his agency to undertake tougher inspection procedures: "[Cherry] puts you in a big fishbowl and looks for all your mistakes. I think this makes for better regulation."[48] His efforts led to the ratcheting-up of inspection standards that NRC inspectors imposed as time went on.

One can only wonder what would have happened if Consumers Power had agreed to fix the hole in Cherry's senior partner's boat. But delays in the construction permit hearings caused by Cherry and tougher scrutiny by government inspectors were not Consumers Power's only problems with the Midland Plant. Bechtel Corporation and the utility's inability to construct the facility to meet the standards of the AEC and the NRC was first noted in 1970 and continued to be a serious recurring problem.

Construction Quality Control

An Unredeemed Promise of Reformation

Where the Midland Plant was built was very important to its proponents, who wanted a low-cost steam supply for Dow Chemical Company, but opponents saw the location as a safety risk. The conflict between Midlanders who were enthusiastic backers of the Midland Plant—the company executives who wanted the plant to successfully advance their companies' goals—and the plant's opponents and their lawyers makes up only part of the Midland Plant story. Another complication developed as construction of the plant proceeded in the 1970s, and the inability of Consumers Power and Bechtel to meet the standards dictated by the Atomic Energy Commission (AEC) and the Nuclear Regulatory Commission (NRC) resulted in negative editorial comments in Michigan newspapers about the very use of nuclear power.

Construction began in June 1969 with site clearing and initial earthwork activities directed primarily toward cooling pond development and excavation in the power block area (location of the auxiliary building, reactor buildings, and turbine building).[1] Because the Midland Plant was not located near a large body of water, it was necessary to construct an 880-acre pond to provide cooling circulating water to remove heat from the steam condensers for the steam turbines, water for a fire protection system, and water for a service water system to cool numerous plant

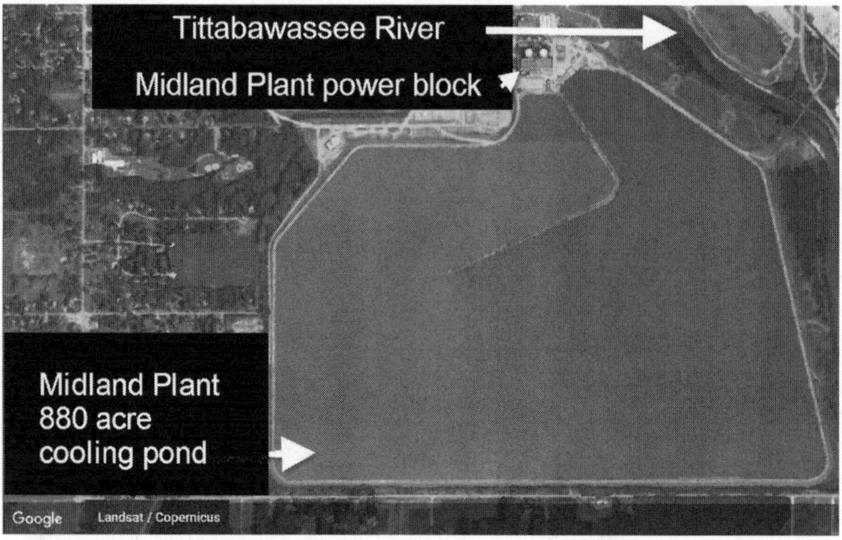

FIGURE 10. Construction of the 880-acre cooling pond and mats for structures in the power block area (auxiliary building, reactor buildings, and steam turbine building) of the Midland Plant began in June 1969. This picture illustrates the large area that was needed for the Midland Plant. (Courtesy of Google Earth/Landsat/Copernicus with notations added by author)

components. Although it was still waiting for its construction permit, Consumers Power was granted a variance to commence limited work in July 1970. Under the terms of this variance, the utility proceeded with concrete work in August, pouring base mats (layers of concrete) for the auxiliary building, reactor buildings, and Unit 1 turbine building.[2] These base mats were poured on top of a layer of hard, impermeable clay that was 135 to 190 feet thick. Beneath the clay was a layer of sand and gravel glacial deposits 160 feet thick that rested on bedrock.[3]

In August 1970, Bruce Peck, a new Consumers Power employee, was assigned to the Midland site. He had hired on with the company prior to completing his service in the U.S. Navy, and a couple of months later, he began working at the Midland Nuclear Plant project. He also moved his family into a house he bought in town.[4]

But three months later in November, Consumers Power stopped all construction activity and laid off all construction workers at the Midland site pending the issuance of the construction permit. That month, Russell Youngdahl, senior vice president for engineering and construction at Consumers Power, said, "It is clear the company is not likely to have an AEC construction permit when it was anticipated, but that most of the site preparation work has been completed and the company already has finished most of the work it planned to do before the permit was issued."[5] At the same time, the company also suspended all engineering work and planning for design revisions incorporating current standards for plant safeguards and control systems.[6] As described earlier, several conservation and citizens' organizations had filed petitions with the AEC to halt construction of the plant until further research had investigated the possibility of health hazards.[7]

Thus, Consumers Power's abrupt stoppage of all site and engineering work left Peck with no apparent work. His expertise was nuclear plants. In college he had completed Navy Reserve Officer's Training, and upon graduation, he had been selected for the navy's eighteen-month Nuclear Power School, the premier nuclear engineering program in the country. After graduation, he spent four years assigned to nuclear submarines. He had accepted his job with Consumers Power with the belief that the utility's ambitious nuclear program was an ideal fit with his training and experience. But now he had been stranded in Midland with no job, a family to feed, and a mortgage he had just signed. But Consumers Power came through with a plan whereby Peck could stay in Midland. They assigned him to construction projects in nearby Flint and Bay City and told him to check the Midland site once each week. When construction restarted, he returned and continued to work at the Midland Plant site for the rest of his career.

Although not directly involved with the concrete installation, Peck remembers the investigation into what was probably the first quality control problem that Bechtel and Consumers Power had with the Midland Plant. To ensure that concrete will form a uniform hardened material when it is poured, equipment is placed into the newly poured concrete to vibrate it and collapse any existing voids.

An AEC inspector on the Midland site in 1970 observed workers not using the vibrating equipment properly while pouring concrete for buildings, including the reactor structures.[8] The inspector's report was sent to Consumers Power. Peck recalled that a company employee then proceeded to write some derogatory comments on it. When the Atomic Safety and Licensing Board held a hearing to deal with the cement matter, Consumers Power had to produce the report it had been sent. Peck believed that these notes probably contributed to the investigating committee's finding that the company did not have a proper concern for quality control. A three-man Atomic Safety and Licensing Board's March 1973 report on the improper concrete pouring included an unusual requirement: that Consumers Power file monthly reports of all construction problems and corrective actions. In September 1973, the appeal board issued its final ruling stating that it did not think quality assurance would be a problem in the future. With this final ruling, the board that had reviewed the concrete problem was disbanded.[9]

It seems worthwhile at this point to spell out the exact 1973 definitions of "quality control" and "quality assurance" that the Atomic Safety and Licensing Board used. These terms were defined by AEC regulations (10 C.F.R., Part 50, App. B): Quality assurance comprises all those planned and systematic actions necessary to provide adequate confidence that a structure, system, or component will perform satisfactorily in service. Quality assurance includes quality control, which comprises those quality assurance actions related to the physical characteristics of a material, structure, component, or system which provides a means to control the quality of the material, structure, component, or system to predetermined requirements.[10]

In early November 1973, almost immediately after disbanding the the three-man concrete review board, AEC inspectors found serious deficiencies in the cadwelding of metal rebar used to strengthen the concrete of the thirteen-foot thick base mats that the hot, highly radioactive reactor core would sit upon.[11] Cadwelding allows the joining of two bars end-to-end and results in a bond of greater tensile strength than individual bars.[12] The discovery of inadequate quality assurance for cadwelding angered the members of the disbanded board. In November 1973 all three of those former board members sent a scathing letter to L. Manning Muntzing, director of regulation for the AEC. In reference to the 1973 welding problems at Midland, their letter said, "What we have here is a pattern of repeated, flagrant and significant QA violations of a non-routine character—coupled with an unredeemed promise of reformation."[13] The former board members revealed they had nearly revoked the Midland construction permit when they learned about the concrete problems of 1970. They hadn't done so because of the "solemn assurance" of the applicant that all of those prior deficiencies were being corrected.[14]

As a result of the discovery of the bad welds and the complaints from the former quality assurance board, the AEC ordered all cadwelding stopped on the Midland Plant on November 9, 1973. The Midland Nuclear Power Committee sprang into action and made a trip on December 10 to Washington, D.C., to get the order lifted. Traveling from Midland to Washington were Dr. Wayne North; Robert (Bud) Parker, director of the Midland Area Chamber of Commerce; Alan Ott, president of Chemical Bank and Trust Company; and Bruce Benway, vice president of the First National Bank and former Chamber of Commerce president.[15] On December 19 cadwelding was allowed to resume.[16]

According to the *Ann Arbor Sun* in 1974, the problem with the cadwelding was dealt with as follows:

> Since construction began, the AEC's regional office in Chicago has been earnestly trying to enforce its complex, stringent regulations. By the time violations reach the agency's higher realms, however, there is still a great deal of sympathy for the Utility's problems, its economic necessities and the status quo. Last December after quality control violations shut down the plant, for example, the AEC's director of regulation, L. Manning Muntzing, met with two Midland bankers, the director of the Midland Chamber of Commerce and a leading minister. The meeting was arranged by Midland's Congressman, Rep. E. A. Cederberg, a ranking Republican on the House Appropriations Committee and a backer of the nuclear plant. According to participants, Muntzing told the Midland businessmen that the hearing necessary to reopen the plant "would go away." Although Muntzing denied the report, what did go away was the order stopping construction, seven days later. AEC inspectors had steered the utility through an announced inspection prior to the meeting, and when the hearing was finally held the next summer, AEC inspectors said there were no longer any quality control problems at the site.[17]

The *Philadelphia Inquirer* did not repeat everything in this account, but it did confirm the meeting and reported much of the same information They reported that Muntzing said that no member of Congress was involved in setting up the December 10 meeting with the Midland Nuclear Power Committee but that a member of Cederberg's staff confirmed that his office had arranged it so members of the Midland committee could urge Muntzing to permit resumption of construction. The newspaper further reported that he said that political pressure played no part in the decision but that a special inspection of the plant on December 7 had resolved the problems.[18]

In spite of the fact that the AEC said specific problems with cement pouring and cadwelding were corrected, these early problems at the Midland Plant prompted negative comments by the commission. In February 1974, James Keppler, the AEC's regional enforcement person in Chicago, had this to say about Consumers Power: "The major problem we have identified was that Consumers was not involved enough in what was going on. They have been relying very heavily on their architect-engineer (Bechtel Corp. of San Francisco, which is building the plant), and they were not getting the job done. The impact we're trying to make on Consumers is that they have to be involved."[19] In the same *Detroit Free Press* article, L. Manning Muntzing, Keppler's boss in Washington, said, "What they [Consumers Power] don't seem to understand is that in a fossil [fuel] plant, if a valve leaks a little bit or pump doesn't pump when it should, it's no big deal . . . In a nuclear plant, we're dealing with very hazardous materials. We're not going to let these things occur."[20] A lot of different quality control problems were noted during the building of the Midland Plant after 1974, and the AEC (and later the NRC) took the point of view that each the of them was "a big deal."

When asked what he remembered about his early experiences with construction at the Midland Plant, Peck recalled that when he brought up problems on the job site to Bechtel, he was told that he should go back to his office and that Bechtel was in charge. His navy experience in nuclear power was with Admiral Hyman Rickover's approach to nuclear power, which was that the navy was the customer and the navy was in charge—contractors needed to march to the navy's orders. But as Peck remembers it, it was not until the late 1970s that Consumers Power management took an active role in construction supervision at Midland.[21]

Another Consumers Power engineer who worked at the Midland Plant site, Bob Wheeler, described Bechtel's response to questions he posed as "why do you keep bothering us."[22] In 1976 press reports described Consumers Power's and Bechtel's construction work as "sloppy" after it was discovered that in twenty-seven instances steel reinforcing bars either were left out of concrete walls or were improperly installed.[23] Quality assurance issues continued to plague the Midland Plant in the late 1970s and early 1980s, and the NRC's Licensing Board said in 1982 that the 1973 and 1976 problems showed that "non-compliance with the Commission's quality assurance regulations is . . . a problem which has plagued the construction of this facility."[24] A 1976 *Detroit Free Press* editorial pointed out, "It is precisely this kind of shoddiness and inability to control human error that makes the nuclear equation so questionable, even if all the safety and environmental concerns inherent in nuclear power were to be convincingly met."[25]

1975

Carry on with the Midland Plant

Construction management in Midland was just one of the three big challenges Consumers Power Company faced. Challenge two was who would be the top person directing the project. Challenge three was how Consumers Power was going to be able to pay for the ever-increasing cost of the Midland Plant. As noted earlier, Jim Campbell died from a heart attack in 1972. An engineer, he thought he could guide Consumers Power's effort to build nuclear generation. With his death a successor was needed to fulfill his nuclear vision. In 1972, rather than make the difficult decision of picking a new president by promoting a vice president or finding a candidate from outside, the board of directors asked A. H. Aymond to assume that position, in addition to being chairman of the board. Aymond turned out to be both inept in managing the funding for construction projects and unable to contain their cost overruns.

In my experience in the 2000s, company annual reports are sunny documents full of reports of great things that have happened in the prior year and glowing predictions of things to come as management guides a company to ever improving financial results. Even if the company had a terrible year, some obscure business metric is trotted out to show that, with the great new programs management will put in place, the future looks bright. However, Consumers Power's annual report for 1974 did not anticipate this trend. In his accompanying letter, Aymond said,

> In my lifetime, there are some years that one would rather forget. And 1974 was one of them. In a word, it was terrible. Just about everything that could go wrong did so. Or, so it seemed. The automobile industry which began the year in a mini-recession as a result of the Arab oil embargo, concluded 1974 in a state of near disaster. Layoffs idled hundreds of thousands of auto workers and shut down a large percentage of the industry's assembly lines.[1]

The Arab oil embargo was a result of the Yom Kippur War. On October 6, 1973, an Arab coalition led by Egypt and Syria launched a surprise attack against Israel in the Sinai Peninsula and the Golan Heights. Then on October 19, retaliating for President Richard Nixon's request to Congress for $2.2 billion in emergency aid to Israel, the Organization of Arab Petroleum Exporting Countries (OAPEC) instituted an oil embargo on the United States.[2] This caused the price of oil in the United States to go from $2.90 to $11.65 a barrel by January 1974. Finally in March, amid disagreements within OAPEC over how long to continue the punishment, the organization officially lifted it.[3] The higher oil prices, however, remained.[4] This spike in oil prices had a devastating effect on the sales of gas-guzzling cars made in Michigan.[5]

What Aymond did not expressly say in his letter to shareholders was that in 1974 Consumers Power's electric sales—which had been growing consistently from 1967 to 1973—declined (1975 would also show a year-on-year decline).[6] The auto economy was in free-fall, and the entire U.S. economy declined dramatically in 1974 with the gross domestic product growth rate going from 5.65 percent in 1973 to a negative .54 percent in 1974.[7] Although Consumers Power lowered its forecasts of future customer need for electricity in 1974, it continued to maintain that the Midland Plant was needed and that the central problem for utilities was raising new capital for new plants so there would not be a serious shortage of electric energy in the late 1970s.[8] Further exacerbating the utility's financial problems in 1974 was its inability to use its largest electric generating facility, the Palisades Plant, which was down for repairs almost the whole year. Aymond called the resulting need to purchase replacement power "a grievous financial burden to the Company."[9]

Because of Consumers Power's decline in earnings in 1974, it could not sell new securities in the amounts needed to continue its major construction projects. Why management had not secured financing for the Midland Plant prior to 1974 is not clear, but the failure to resolve this funding problem did not speak well for Aymond's financial management abilities. To deal with its 1974 financial problems, the company cancelled its plans to construct the $1.4 billion twin unit Quanicassee nuclear plant east of Bay City and severely reduced the construction funds for the Midland Plant in 1975.[10] With this cutback, the estimated in-service dates for the two reactors there—which had originally been 1974 and 1975—had to be pushed back to 1981 and 1982.[11]

A problem beyond the control of Consumers Power's management was the U.S. economy. The surprising thing about its decline in 1974 was that it did not cause a decline in the rate of inflation. In 1973 Americans were on a spending binge. The Federal Reserve confidential "red book," a discussion of regional conditions, described on October 10, 1973, a very strong economy with widespread shortages of manpower and materials causing prices to rise as dollars chased smaller supplies of goods. With the booming economy, businesses raised prices and inflation escalated. Although the rise in oil prices killed economic growth, inflation remained along with high interest rates. This persistent inflation along with economic stagnation took on a name of its own: stagflation.[12]

Due to the inflation part of stagflation, the interest rate Consumers Power had to pay on the funds it was able to borrow was rising sharply in the last half of 1974. The rate had come down slightly from the Vietnam War inflationary rate of 8.6 percent in 1970 to 7.4 percent in November 1971.[13] But in February 1974, when the utility sold $60 million in first mortgage bonds, the market required it to pay an effective rate of 11.6 percent, and in September 1974 a second issuance of first mortgage bonds was sold at an effective rate of 11.4 percent.[14] As noted earlier, when Consumers Power announced the Midland Plant in 1967 it had been able to sell first mortgage bonds at an interest rate of 5.9 percent. This change of over 5 percent in borrowing costs in seven years meant that for every $100 million the company was borrowing by 1974, it was paying $5 million more per year than it had paid in 1967.

In April 1975, Consumers Power announced that the estimated cost of the Midland Plant would be $1.4 billion.[15] The previous year it had announced an estimated cost of $940 million, saying that inflation in the cost of labor and materials, changes in plant design as a result of frequently changing Atomic Energy Commission (AEC) construction guidelines, and delays caused by environmental groups opposing the plant were all responsible for the increased cost estimate.[16] The only new reason Consumers Power offered for the $1.4 billion estimate was that when the company was pressed for cash in 1974 and had to cut back on construction, that delay had

TABLE 2. Estimated cost to complete the Midland Plant made by Consumers Power by year.

YEAR	ESTIMATED COST TO COMPLETE
1967	$267 million*
1969	$375 million†
1971	$500 million‡
1974	$940 million§
1975	$1.4 billion‖

* "Consumers to Build $267 Million Nuclear Plant for Dow Power Use," *Midland Daily News*, December 14, 1967, 1.
† Betty Herring, "Nuclear Plant Called 'Greatest Investment in the State,'" *Midland Daily News*, June 27, 1969, 1.
‡ Jerry M. Flint, "Midland Mich. Wants Nuclear Power," *New York Times*, December 3, 1971, 24.
§ "Costs Are Soaring For Power Plants," UPI, Holland (MI) *Evening Standard*, January 9, 1974, 14.
‖ "Midland N-Plant Cost Jumps 48%," *Midland Daily News*, April 12, 1975, 1.

cost $232 million.[17] The utility had regularly increased the estimated cost from the original $267 million, but this 1975 jump to $1.4 billion was over five times their original estimate. The history of the estimated cost to complete is summarized in table 2.

▪ ▪ ▪

In addition to the 1975 huge escalation of the estimated cost to complete the Midland Plant, key decisions Consumers Power had made on other projects were creating major problems. One of these involved the Palisades Plant, where the selection of a contractor that had never built a steam supply system for a nuclear plant was turning out badly, with the facility down for repairs almost the whole year.[18] One of Consumers Power's solutions to this problem was to become mired in a lawsuit against the equipment and engineering companies (which included two Bechtel Group companies) that had supplied engineering services and equipment for the plant.[19] Another decision involved its Marysville synthetic natural gas plant, completed in 1973 at a cost four times the original estimate.[20] The 1974 estimated cost of gas produced from the plant was five times what the utility was paying for other gas because of the cost of petroleum feed stocks the Marysville plant required.[21]

Although the Michigan Public Service Commission (MPSC) grudgingly allowed Consumers Power to pass through to its customers all of the construction cost of the Marysville plant as well as the cost of gas produced there, the company complained about the fact that the commission's approval of the Marysville's cost carried a rebuke warning the utility's board of directors to do something about cost overruns.[22] Consumers Power's need to build the plant arose from a mistake it had made years earlier when (unlike the other big gas utility in the state, Michigan Consolidated) the utility had become totally tied to one interstate pipeline company, Panhandle Eastern Corporation, for the supplies of natural gas from the southern United States. When Panhandle cut Consumers Power's supplies, the only remedy the company could find was to build this synthetic gas plant, and the MPSC was put in the position of going along with it or possibly facing a situation in which Michigan customers did not have enough gas to heat their homes in the winter.[23]

That wasn't the only problem with the Marysville facility. Consumers Power decided to make it ten times larger than synthetic gas plants that had been previously built with a similar design in England. When engineers scaled up from these early plant designs, they found the huge Marysville plant was extremely more complicated than they had thought. Original estimates called for one million man-hours of work, but with the added complexity, man-hours ballooned to over four million.[24]

Aymond told the *Detroit Free Press* in June 1974 that the State of Michigan had no choice but to give the company what it needed—meaning rate increases—or else it would stop its entire expansion program. "The public," he said, "has to keep us viable, whether or not some of our costs were prudent."[25] This statement may sound so extreme, one might wonder if he was misquoted, but long-time Consumers Power employees in the late 1980s said their company's position on rate increases in the 1970s makes Aymond's statement seem plausible.[26]

Aymond also liked to pick fights with Michigan's attorney general, Frank Kelley. Aymond said in June 1974, "No agency has taken a position as contrary to the public interest as that of Frank Kelley. I don't see why there should be an adversary when there's no legitimate opposition to our request."[27] A frequent critic of the utility, Kelley had legitimate problems with Consumers Power's constant requests for increasing electric rates and its choice of generation facilities. In 1972, the attorney general forced the company to settle for an increase of $28 million in electric rates, half the $56 million requested.[28] Then in February 1974, Kelley said that the major portion of the 6 percent increase in Consumers Power's electric rates was necessitated by the company's failure to operate and construct modern, efficient generation units.[29] Aymond's general attitude was that those people in Lansing—the attorney general and the MPSC—had no idea what needed to be done to supply electricity to Consumers Power's customers.

Given Aymond's adversarial relationships with the attorney general and the MPSC, both of whose approval the company needed to raise its rates, and his inability to manage the huge increases in costs of the Marysville Gas Plant and the Midland Plant, it is not surprising that in 1974 Consumers Power was looking to make changes. In July 1974 Aymond told the *Detroit Free Press*, "I would hate to leave the company where I've spent my life under a cloud of incompetence." But he conceded that it could happen.[30]

▪ ▪ ▪

By 1975 utilities all over the United States were canceling nuclear plants because of the same things that Consumers Power was experiencing: a decline in electric demand, the rising costs of the plants, and high interest rates. Utilities ordered 231 nuclear plants from 1958 through 1974, but only fifteen in the period 1974 to 1982. More than one hundred nuclear plants were cancelled between 1974 and 1982, even several that were well under construction.[31]

With the advantage of over forty-five years of hindsight, the case can be made that in 1975 Consumers Power should have fired Aymond and cancelled the Midland Plant. Under this scenario the utility would have claimed that the escalation of labor and materials cost, higher interest rates, and lower electric demand changed the situation radically from when it had announced the plant in 1967. Although it had $205 million invested in the Midland Plant by this time, the MPSC starting in 1973 had allowed both Consumers Power and Detroit Edison to recover from their customers expenses they had incurred in nuclear plants that the companies cancelled (Fermi 3 in 1973, Quanicassee in 1974, and Greenwood in 1980).[32] It would have been reasonable to assume that the commission would have allowed Consumers Power to do the same with Midland in 1975. But if the board of directors discussed using the replacement of Aymond as part of an off-ramp to cancel the Midland Plant, there seems to be no record of it.

What actually happened was that Consumers Power's board of directors gave the titles of president and chief operating officer to John Selby effective April 1, 1975.[33] The hiring of Selby—who was called an electronic and nuclear engineer in the utility's 1975 annual report—from General Electric's energy products division was a firm indication that the company intended to "carry on" with the construction at Midland without regard to what was happening in the rest of the utility industry. And in spite of his lackluster performance as president, Aymond retained the title of chairman of the board.

The selection of Selby deviated from the normal Consumers Power's practice of promoting from within. Before coming to work for the Michigan utility, he had spent twenty-eight years with General Electric. His last job there had been deputy division general manager for its nuclear energy products division, the unit that made nuclear fuel and nuclear steam supply systems for generating plants.[34] Although he had been most recently working in San Diego, Selby told the *Detroit Free Press* in 1975 that he was glad to move to Jackson, Michigan. A Midwesterner, he grew up in the small Iowa town of Odebolt—a place that called itself the popcorn center of the world—and his general engineering degree was from Iowa State. He said he wanted to run a company, and he couldn't do that at General Electric. At fifty-four, he had gone past the age that he could become chairman there.[35]

Selby must have had strong confidence in his abilities to run Consumers Power. In taking the job in 1975, he was joining a company with construction quality and cost control problems at its Midland Plant, an adversarial relationship with the MPSC and the Michigan attorney general, a dispute with Bechtel Corporation, and problems with a history of needing to continually renegotiate its contract for energy supply with the Dow Chemical Company.

■ ■ ■

Due to the problems Consumers Power encountered with the Palisades Plant, the utility sued Bechtel and three other suppliers for $300 million in 1974. The obvious question was if Bechtel didn't do a satisfactory job at the Palisades Plant, why was Consumers Power using the engineering firm to build the Midland Plant?[36] In 1977 Consumers Power settled the suit with Bechtel for a payment of $6 million in cash and $8 million in future services. A Consumers Power spokesman said, "Settling the Bechtel suit makes it easier for us to work with them at Midland . . . We've been subjected to a tremendous amount of criticism for using them at Midland while we sued them at Palisades."[37] In 1979 an executive at the utility said, "I look upon Bechtel as being one of the best construction firms in the country."[38] So it seems that Selby might have believed when he took the job that he could make the Bechtel problem go away with a legal settlement.

Whatever Selby's plans were to solve the problems with the Dow Chemical–Consumers Power energy contract, he ran into continuing problems. The contract between the two firms was renegotiated in February 1974, after being extended eighteen times.[39] The terms were not disclosed at the time, but we now know that the new agreement gave Dow Chemical the flexibility to buy only half as much steam as in its original 1967 agreement and to keep operating its own generating plants. Dow Chemical apparently also committed to pay four times as much for its steam as it had originally anticipated because of the huge increase in the cost of the Midland Plant and a sizeable though less dramatic increase in the cost of nuclear fuel.[40] It became obvious that the February 1974 new contract was not working for Dow Chemical when in November the chemical company expressed "great disappointment" that Consumers Power had once again moved back the date they could supply steam.[41] By 1976 Dow Chemical's attorney Leslie Nute would call Consumers Power's efforts to keep the contract in place "pretty damned close to blackmail."

■ ■ ■

Prospects for the utility and its new president Selby did improve in 1975 as Consumers Power was able to start collecting increased electric rates authorized by the MPSC that January. In addition, due to a new oil-fired Unit 3 at the Dan E. Karn Generating Plant, the company was able to significantly reduce the need for high-cost purchased power. As a result of its improved financial situation, it was able to issue new securities to help finance its construction program.[42] It seemed at the end of 1976 that Consumers Power had withstood the shock of the Arab Oil embargo and that Selby was going to lead the company forward free, they hoped, from new quality control issues and new adverse events from outside Michigan.

At the Midland Plant, the curtailment of engineering work and the reduction in procurement and construction imposed for all of 1975 due to Consumers Power's lack of cash were reversed in 1976, and construction progress resumed at the anticipated rate. The drastic cutback in the workforce ended in 1976, and it was announced in November that the total craft workforce would double to 1,600 in the next calendar year to keep pace with the workload.[43] Like many other dates and numbers announced for the complex Midland Plant with its two large reactors and steam supply system for Dow Chemical, the number of employees would prove to be a moving target and further demonstrate Consumers Power's management challenges.

CHAPTER 15

Midland Plant

Opportunities (and Frustrations) for All

The 1976 pronouncement that 1600 craft workers would be working at the Midland Plant by 1977 was jaw-dropping compared to Consumers Power Company project engineer Willian E. Kessler's 1969 prediction that a workforce of between five hundred and seven hundred would be employed at the peak of construction.[1] But even the 1976 estimate would prove to be far off the mark. In January 1978 project superintendent Thomas C. Cooke reported 2,750 workers on the site, 2,200 in manual labor and 550 in nonmanual jobs, in addition to seventy subcontractors and fifty Consumers Power testing employees.[2] A 1978 Labor Day salute to workers at the Midland site included laborers, bricklayers and cement masons, boilermakers, carpenters and millwrights, painters, pipefitters and pipe welders, ironworkers (rebar, structural, and riggers), operating engineers, teamsters, and electricians.[3] At year's end, 4,200 engineers and construction people were working on the Midland Plant.[4] By July 1983 this workforce would grow to approximately 5,500 people.[5]

Having all of these workers together at one job site created a unique environment. One Dow Chemical Company employee described the Midland Plant jobsite as a bazaar where all sorts of things were available for purchase from plant workers who were running their own small businesses. He said, "Every day was a huge bake sale with all manner of cakes and cookies for sale."[6] Whereas at the Dow Chemical and Dow Corning facilities the workers enjoyed spotless company cafeterias for their lunch, at the Midland Plant site a pipefitter sold chili he made daily in the space under the containment building.[7] With all kinds of high-quality metal available, a few workers manufactured special items for sale, such as handcrafted belt buckles.

Some things that may or may not have been available there are hard to document. The buildings and construction trailers contained many rooms. According to several reports, a madam took up residence in one of these rooms to provide services to the workers. This story, however, may be a legend, as none of my sources wanted to be quoted as having specific knowledge of a Midland Plant brothel. A story that I was able to substantiate, however, was the creation of scatological "artwork" and "poetry" at various locations. Although the artist or artists were never identified, graffiti was regularly exhibited on stairway walls and restrooms stalls. Its appearance so infuriated a Bechtel Corporation supervisor that he assigned an employee to tour the site every morning with a can of black spray paint to obliterate the latest postings of the person dubbed the "shit house" poet. According to Bruce Peck, at a fall 1979 dinner and roast for departing Bechtel construction site manager Joel Newgen, a special award was conferred: a restroom stall door decorated with graffiti.

Off the Midland Plant job site, what did people in town think about these thousands of new workers? One doctor who was practicing general medicine during the time of construction remembers that area bars started opening before work in the morning instead of in the afternoons and evenings, and that some Bechtel workers would stop by Frick's Bar and have a couple of shots before going to work. One woman who worked for Dow Chemical at this time said that she and her colleagues were shocked when one of their married coworkers announced that the after-hours opportunities in the local bars were so improved that she had to get several new sets of lingerie.

Another person said that, with the Bechtel workers in town, the streets became what many referred to as the Bechtel 500. Others mentioned the new practice of turning right from the lane next to the right-hand turn lane. This practice may have been the norm in places like California, but in Midland if you wanted to turn right you did it from the far right lane. Turning right from the lane next to the right hand turn lane became "the Bechtel turn."

As for the workers themselves, the Midland Plant provided many good job opportunities. In 1976 Bechtel hired Linda Lein (later Linda Sims), a new journalism graduate from Michigan State University, to be the editor for a jobsite newspaper.

FIGURE 11. November 1976, Linda Lein was hired as first editor of the *Midland Reactor*. (Courtesy of Linda [Lein] Sims)

She said that when she interviewed for the job, she was somewhat ambivalent about nuclear power, but a job where you could wear Levi's to work had a lot of appeal for a young woman looking for her first job. A contest to name the newspaper was held, and the winning entry, submitted by laborer Jack Groeschen, was the *Midland Reactor*.[8] In her initial edition in November 1976 she wrote, "The *Midland Reactor's* purpose is to serve as a source of information on current developments on the jobsite, technical as well as personal. While significant developments on plant construction will be reported, so will efforts on a personal level." Lein left Bechtel after two years for a job with Consumers Power, but she set the standard for a publication that recorded workers' progress and was distributed to all Midland Project employees—Bechtel, building trades, Consumers Power, and subcontractors.

That inaugural issue of the *Midland Reactor* included a picture of the eighteen graduates (White and all male) of an onsite welding class. Under the picture was an article titled "Wanted: Qualified Welders to Fill the Site Demand for the Increasing Work Load at the Midland Project." The article continued, "In order to fill this need at a time of nationwide shortage of qualified welders in the mechanical construction trades, courses are being offered to train and upgrade welding skills of employees through free, onsite programs and Adult Education through the Midland Public School System."[9]

But since times were changing in the 1970s, opportunities also existed for those who were not male and not White. The *Midland Reactor* profiled Liz Uptegraff (called the "queen bee" of the second shift) in February 1977. She was a pipefitter apprentice and was receiving on-the-job training at the job site while also taking classes. It also noted that Uptegraff, the first female pipefitter apprentice on site, was also the first lady in the Union Local 85. She told the *Midland Reactor*, "Sure,

I'm an oddity, being a woman construction employee, but the guys treat me as an equal and as a good friend."[10]

After Midland, she went on to a career as a welder and worked on other nuclear plants: the Vogal plant in Georgia and the Nine Mile Point plant in New York. She has a vivid memory of the extremely close tolerances that were demanded on all of the welding on nuclear plants. At Midland, her work was checked by Bechtel and then often rechecked by Consumers Power. She takes great pride in the work she did at Midland and the other nuclear plants. All of her work had to include a unique number assigned to her, and she feels that she established a portfolio as an artist would. She felt a great sense of loss when Consumers Power, to be able to declare a tax loss, deliberately cut the line that she had helped construct to move steam to Dow Chemical.[11]

One of the workers featured in the January 1981 issue of the *Midland Reactor* was a young Black man, Dana Crew, who was a first-year millwright apprentice. "I like working at the Midland Project. The work is interesting and varied, and people are great," he said. He was dividing his time between Midland and Detroit, where he had previously been a repair relief man in an automobile factory.[12]

Chris Gallandt, an electrician apprentice, also was profiled in the newspaper. "After receiving a degree in history, she found the job market tight. Her father, also an electrician, suggested this apprentice program. Embarking on another four years of education, Chris is involved in a whole new kind of learning including physical skills she has not previously used."[13] She proceeded to become a journeyman electrician. She also noted that as she remembers her days on the Midland job site, of the one thousand electricians she was the only woman. She said that there were so few women there that there were only two restrooms for them. One was in the office, which she felt uncomfortable using. The second was on the far end of the site, a fifteen-minute walk.[14]

Gallant said the reason there were apprentices at the Midland Plant was the union required that apprentice jobs be available. A 1980 *Midland Reactor* article detailed the agreement: "The Special Project Labor Agreement contains special provisions for implementing the full utilization of apprentices." The article stated that the opportunities for minorities to receive on-the-job training were reflected in the increasing numbers of minorities and women in apprentice programs at the Midland site. At the time, one hundred apprentice workers in eight crafts were assigned to the Midland Project.[15]

However, not all of the job opportunities at the Midland Plant were in the construction trades. In her profile in the *Midland Reactor*, Pegg Mudd (now Pegg Woodby), said she started in 1973 as a switchboard operator working for Bechtel.[16] In 1974 when Bechtel required her to go out on the job site to pick up drawings, she was the first woman allowed to walk the site, and she received detailed instructions before she was allowed to go out. The instructions included a firm warning that she was to ignore all catcalls. Following her job with Bechtel, she went to work for Consumers Power at the Midland Plant site in 1976, processing plant documents. At the time of the article in the 1982 *Midland Reactor*, she had advanced to a job in which she supervised a staff of seventeen, and her group was starting on a two-year project to microfilm seven million pages of documents.

▪ ▪ ▪

Besides opportunity and pride, there was another constant theme in the workers' stories—frustration. Quality control problems and changes in regulatory requirements caused a continuing need to tear out meticulously constructed electric wiring and plumbing, as well as to take down all sorts of structures. Although there seems to be no specific documentation detailing the cost of change orders and the time required for their implementation, through almost all of the history of construction on the Midland Plant, the utility complained about the constant changes in Atomic Energy Commission (AEC) and later the Nuclear Regulatory Commission (NRC) design specifications. In a 1974 briefing to the Midland Chamber of Commerce, Russell Youngdahl explained that one of the factors driving up the cost of the Midland Plant was the changes in plant design required by these regulatory updates.[17] Nine years later in 1983 John Selby summarized the company's experience: "We have had a moving target. The whole industry has. We are not unique. But I think that a plant that was basically designed for the requirements, the codes and standards of the late 1960s to meet the requirements, codes, standards, safety requirements in the 1980s is a significant undertaking, and a lot of that has been the situation in Midland."[18]

It wasn't just the craftspeople who suffered from the frustration caused by the constant changes, nor was Consumers Power's experience unique. A 1980 Oak Ridge National Laboratory report on nuclear power provided an estimate of increases in material and labor caused by the number of standards and regulations affecting nuclear power plant design issued by the American National Standards Institute, the AEC, the NRC, the Environmental Protection Administration, the Occupational Safety and Health Administration, and other governmental agencies.[19] Estimated percentage increases from the early to late 1970s can be seen in table 3.

TABLE 3. Percent increase in costs of constructing nuclear power plants from the early to late 1970s.

MATERIAL/LABOR	INCREASE
Structural steel (pounds/kWe)*	43%
Reinforcing steel (pounds/kWe)	39%
Piping (lineal foot/kWe)	50%
Electric cable (lineal foot/kWe)	37%
Manual and nonmanual labor (hours/kWe)	50%

*kWe means kilowatt of electricity and is used because the plants studied ranged in total generating capacity.
Irving Spiewak and David F. Cope, *Overview Paper on Nuclear Power* (Oak Ridge, TN: Oak Ridge National Laboratory Report ORNL/TM-7425, September 1980), 108–109, https://inis.iaea.org/collection/NCLCollectionStore/_Public/12/582/12582196.pdf?r=1&r=1.

Physicist Dr. Bernard Cohen, an advocate of the view that it was safe for humans to be exposed to low-level background radiation, described the regulatory changes that occurred during the period of the Midland Plant's construction as "ratcheting." He noted that in response to escalating public concern, the NRC began implementing more regulations in the early 1970s and continued to issue more throughout the decade. The regulatory requirements were constantly tightened, requiring additional equipment, construction labor, and materials.[20] In response to the continuing need to tear things out and then build them over again and to show their empathy for their craft workers, Consumers Power management issued this apology to Midland Plant workers in the January 1981 *Midland Reactor*:

> Last year we faced an unusually high number of changed NRC requirements which caused rework in many sections of the plant, particularly in the auxiliary building. It must be frustrating for crafts to complete construction of a system or component only to find that a couple of weeks later it has to be torn out. We are building the Midland Plant during a period of changing regulatory requirements. If the NRC develops new rules, we have to modify work to meet these new rules. All too often in 1980 that meant tearing out a system, redesigning and subsequently reconstructing it.[21]

▪ ▪ ▪

As the number of employees at the Midland Plant site continued to grow, the utility's executives made positive statements about the skills of their workers. Implicit in these statements was their acceptance of the huge increase in the number of workers present at the Midland site over the what they had originally anticipated. In a letter to shareholders dated February 10, 1978, President John D. Selby and Chairman of the Board A. H. Aymond, wrote, "In 1978 we will continue to act to move your Company forward, and that with the continued support of shareholders and the demonstrated skills of our employees, our action will continue to produce positive results."[22] Unfortunately, that year a major construction problem would be discovered, and the ability of the employees to produce "positive results" became extremely difficult.

1978

Not Only Optimism Sinks and Cracks

Consumers Power Company's 1977 annual report, in a grand effort to stay positive about the Midland Plant, said construction had stayed on schedule throughout 1977, projected that Unit 2 would be in service by 1981 and Unit 1 by 1982 (enabling the utility finally to produce steam for Dow Chemical Company), and noted that the estimated cost had not changed from the $1.67 billion reported a year earlier.[1] Also of interest was the mention of the first birthday of Consumers Power's wholly owned subsidiary, Plateau Resources Limited, which had been established to help meet the company's long-term need for uranium to fuel its nuclear plants. At the end of 1977, Plateau owned mining claims covering 122,000 acres in Utah, Colorado, and Wyoming and had options to purchase claims on an additional three thousand acres.[2] It would be reported later that as of December 31, 1978, Consumers Power had invested $34 million in Plateau.[3]

The 1977 annual report mentioned that the construction permit for the Midland Plant was still the subject of litigation, but it had not delayed construction that year. The company reported that it had requested the U.S. Supreme Court to review the 1976 Court of Appeals decision remanding the Midland Plant's construction permit to the Advisory Committee on Reactor Safeguards (ACRS) for review.[4] The implication was that the Supreme Court would reverse the decision of the Court of Appeals, which it did on April 3, 1978, upholding the validity of the construction permit.

This optimistic report, however, did not specifically mention that the estimated cost of $1.67 billion was an increase from the $1.4 billion estimate reported in 1975, nor that in 1977 uranium prices had risen to $22.14 per pound (in 1967 dollars) from the 1967 price of $8.00 per pound.[5] Therefore it seems plausible that the reason Consumers Power found it necessary to enter the uranium mining business, an endeavor in which the company had no experience, was to try to provide some certainty in the fuel's future price.

Another piece of negative news that went unreported was that quality control had once more been a problem. Consumers Power reported on April 19, 1977, that two "tendon sheaths" had been strung in the wrong place for the containment structure of the Unit 1 reactor.[6] Consumers Power fixed this problem. In May 1977 Nuclear Regulatory Commission (NRC) officials from two regional offices and Washington, D.C., visited the Midland Plant for a special inspection of the quality assurance program. They found that the program had been improved and gave Consumers Power and Bechtel Corporation permission to proceed with construction.[7] So as 1978 got underway, once again, the two firms appeared to have their quality control problems solved.

Compared to Consumers Power's original announcement in December 1967, as of February 1978 the Midland Plant's projected in-service date was approximately seven years behind schedule and the projected completion cost 625 percent over budget. In addition, the utility had wandered off into the uranium business where it had no corporate experience. Nonetheless, Consumers Power's story continued to be that Michigan needed the electric energy to meet the increasing demand in the 1980s.[8] The future, however, didn't turn out to match their projections. In 1978 the utility's electric sales were 26.4 billion kilowatt-hours, and in 1984 it sold 26.9 billion kilowatt-hours for an increase in sales of only 1.9 percent.[9] In 1983 Consumers Power's critics took great delight in referencing the company's 1974 prediction of electric demand growth of 7 percent annually.[10]

And the optimistic completion and cost forecasts of early 1978 proved to be just forecasts—they did not foretell what would really happen. On the morning of September 7, 1978, a new example of Bechtel's and Consumers Power's failure to adequately supervise the quality of the construction on the Midland Plant was discovered. This quality assurance problem would result in a new public hearing and require Consumers Power to spend significant money and time to correct it.

FIGURE 12. Emergency generator building is circled. The car in the parking lot in the lower left hand corner of the picture shows the size of the buildings in the picture. Immediately behind the emergency generator (center), the large rectangular building is the steam turbine building. The two silo-shaped structures at the top of the picture are the containment structures for the nuclear reactors. (Courtesy of Google Earth/Landsat/Copernicus with author's notation)

■ ■ ■

Small yellow canaries were once taken into coal mines to detect the presence of bad air. A huge, drab, gray building resembling a concrete bunker was the canary for the Midland Plant. When I saw this bunker-like hulk for the first time, my impression was that behind the four massive doors of the building were four armored personnel carriers that could roar out and secure the site of the Midland Plant. But its real purpose was to house four electric generators, each powered by a diesel engine—this building was called the emergency generator building. When installed, each generator unit included an 84-ton V-12 engine, a 3-ton flywheel, a 29-ton generator, and a 21-ton auxiliary sub-base for a total weight of 137 tons.[11]

On the morning of September 7, 1978, during a routine check of this emergency generator building, Bechtel workers found a dead canary: the building was sinking.[12]

In November Consumers Power said the building had sunk about three inches, as much settling as the company expected over the next forty years.[13] This was particularly alarming because at the time the sinking was discovered, the four 137-ton generators had not yet been installed.

The emergency generator building was classified as a safety-related component by the NRC. This rating is based on the fact that a fundamental safeguard for a nuclear reactor is the ability to remove heat from the reactor's core. The removal of heat in the Midland Plant design was to be accomplished by the circulation of water with pumps powered by electricity and controlled by a complex system of electric circuits. Electricity to run the pump system under normal operations comes from the generators run by the steam turbines that receive steam produced by heat from the nuclear reactor. In the event that this energy is not available because the plant's steam turbines are not running, electricity to run the cooling pumps can flow from outside the plant. But in the case when no electricity is available, the emergency generators are turned on to send electricity to run the cooling water pumps and other safety equipment.

The removal of heat is necessary even when the reactor has been shut down because after the chain reaction is stopped, enough residual radioactive material can exist to generate dangerous levels of heat. If this heat is not removed, a meltdown of the core can result, breaking through the reactor's containment building and releasing radioactive material. Opponents of nuclear power called the meltdown event the "China syndrome," to convey the image that if the molten core of a nuclear plant breached its containment, it would release so much heat and radioactivity that it could penetrate all the way to China. In addition to the possibility of this kind of catastrophe, the heat from a melting core can melt the material surrounding the fuel pellets (cladding) in the reactor core and produce highly explosive hydrogen. In the presence of water, the oxidation of the zirconium in the cladding is accompanied by the release of hydrogen gas.

What would actually happen if emergency generators did not produce electricity when they were needed? Many years after the Midland Plant was being built, the Japanese Fukushima Daiichi Power Station nuclear disaster provided a dramatic example. This site is located in northeastern Japan in a seismically active area. On March 11, 2011, nuclear reactors 1, 2, and 3 automatically shut down due to a magnitude 9.0 earthquake that occurred at 2:46 p.m. With the reactors shut down, no steam was available to power on-site steam turbines to produce electricity. The earthquake severed power lines from outside the plant, and therefore outside power was not available to run the cooling-water pumps to remove the residual heat from the nuclear reactors. The on-site emergency generators had flooded out after the earthquake because of a tsunami that followed at 3:27 p.m.[14] Lacking electricity, the cores in reactors 1, 2, and 3 melted down.[15] The resulting high temperatures generated both vaporized water and a large amount of hydrogen. Hydrogen explosions erupted at all three units, discharging large amounts of radioactive material into the environment.[16] The reactor building of Unit 4, which was not running at the time of the earthquake and tsunami, was also demolished by a hydrogen explosion due to hydrogen leaked in from Unit 3.[17] This explosion exposed the spent fuel storage pool to the atmosphere.[18] In response to the Fukushima accident, an order was issued to evacuate a twelve-mile radius around the site. An estimated one hundred and fifty thousand people left the area, and as of 2019, the estimated cost of the cleanup was $200 billion.[19]

Given what would later happen in 2011 at Fukushima, the NRC's 1978 order to Consumers Power to fix the sinking of the emergency generator building and thus eliminate the possibility of broken connections between the emergency generator building and the rest of the plant was more than reasonable. The first step was to determine why the emergency generator building was sinking. The assessment in a November 1, 1978, NRC memorandum identified four possible causes: (1) compressibility of the fill dirt used to support the building, (2) the design and foundation of the building, (3) piping underlying part of the building, and (4) soil compaction from the filling of the cooling pond.[20] The NRC quickly decided that the reason for the settling of the emergency generator building was insufficiently compacted fill dirt, and this conclusion was documented in a 1981 NRC Circular.[21]

Why would structures for what started out to be a multimillion dollar project (and turned into a multibillion project) be built on fill dirt? As Rod Wieland, a civil engineer who worked on the Midland Plant put it, the facility was "not located at a natural site."[22] No large body of water was nearby to provide emergency cooling water and necessary cool water for the steam turbines, so a huge cooling pond requiring 880 acres of land was constructed. The Midland Plant also needed to be close to the Dow Chemical complex so that a minimum of heat would be lost in transporting steam. The site selected on the floodplain of the Tittabawassee River satisfied both the criterion of space and of proximity, but it had an elevation of only approximately six hundred feet mean sea level (msl). This meant that fill dirt needed to be added to build a hill that could keep the Midland Plant site dry when the Tittabawassee River flooded.

But how much fill dirt needed to be added to raise the level of the Midland Plant site? The decision to add thirty-four feet was based on a study that showed that that the possible "domino" effect of the failure of upstream dams and precipitation could result in a wave-free water level of 631 msl flooding the site.[23] Trucking in thirty-four feet of fill dirt brought the site to an elevation of 634 feet msl, which provided three feet of freeboard against the wave-free flood level.[24] Prior to the building of thc Midland Plant, the maximum historical flood at the Midland gauge station was 610.0 feet msl recorded in 1916.[25] As of May 20, 2020, the maximum flood experienced at the site was approximately 614.20 feet msl, a level that occurred after the failure of upstream dams on the Tittabawassee River.[26]

Given the need for an island constructed of fill dirt, what decisions did Consumers Power and Bechtel make for the foundations of structures at the Midland Plant? For the first group of structures, the reactor containment buildings and most of the auxiliary building, work started before fill dirt was added to the site. Holes were dug approximately nine to thirty-two feet into the upper portion of an approximately 150-foot thick undisturbed glacial clay deposit.[27] Concrete was then poured into these holes to construct reinforced concrete mat foundations.[28] Fill dirt was then placed around these structures. But a second group of structures, including the emergency generator building and a number of other safety-related structures, were built directly on the fill dirt.

To ensure that part or all of the fill will not sink under the weight of the structure built on top of it, a number of steps must be taken. The fill material must be carefully selected and then placement of the fill should be supervised by a qualified geo-technical engineer. Next, the fill needs to be compacted by running equipment over it to make sure that later settling of the fill will not take place. Finally the compaction of the fill dirt needs to be satisfactorily tested. Testimony in hearings held by the Atomic Licensing Board in July 1981 and by the U.S. House of Representatives Subcommittee on Energy and the Environment in 1983 detailed how Consumers Power and Bechtel failed to ensure the proper implementation of the requirements for the satisfactory placement of fill dirt.

Bechtel and Consumers Power's failures became obvious when an investigation of other structures revealed that fill dirt problems also endangered the foundations of other safety related structures, including the borated water storage tank, essential service water intake structure, plant area utilities, and portions of the auxiliary building feedwater isolation valve pits.[29] The borated water storage tanks would hold neutron-absorbing boric acid, which in the case of an emergency could flood the reactors; the service water intake structure held the pumps that would suck cooling water from the pond; buried pipe from the service water building was necessary to carry water for critical cooling functions; and the auxiliary building housed the control room and cabling systems that were necessary to control the nuclear reactors.[30]

Bechtel and Consumers Power's inattention to quality control for buildings' foundations became apparent in the Atomic Licensing Board hearing in July 1981. Gilbert S. Keeley, Consumers Power project manager, said in his written testimony that problems with the administrative building settling unevenly had been discovered in August 1977. When he was asked in the hearing why the errors discovered here, in a building that was not safety-related and thus not under the NRC's jurisdiction, were not corrected by the time work began on the diesel generator building, he testified, "that was due . . . to people not paying attention to detail. Also, there was not a qualified geo-technical engineer on site 100 percent of the time directing these (soil placement) activities." Keeley, who had been director of quality assurance programs at the plant until 1975, also testified that there had been improvements in quality assurance since that time. He said those included hiring more quality assurance personnel, a reorganization of the entire program under which Consumers Power took more responsibility, and the implementation of over-inspection activities. "I think QA is much better now than when I was QA director," Keeley said.[31]

In a suit filed against Consumers Power on July 14, 1983, Dow Chemical would allege that in 1977 the administration building was constructed on the defective fill and excessive settlement occurred immediately; that by August 1977, portions of the structure's foundation had settled up to 3½ inches in one month; and that the investigation conducted by Consumers Power in 1977 found "voids" or pockets of air in the fill and unbroken lumps of clay up to three feet in diameter, demonstrating that the fill had not been compacted, contrary to Safety Analysis Reports certified by Consumers Power to the NRC. At least by the end of December 1977, Dow Chemical claimed, the senior management of Consumers Power knew or should have known that the fill on which the Midland Plant Nuclear Facility was being constructed was patently inadequate for construction of a nuclear power plant and potentially unsafe; and yet Consumers Power decided to proceed with construction of the diesel generator building and other safety-related structures on the fill.[32]

Consumers Power and Bechtel's fix for the emergency generator building was to put 2,000 tons of sand in and around it to compress the fill to maximum compaction.[33] Other proposed fixes were relatively straightforward although they would all take time and money: for the borated water tank foundations, they would build new ring foundations outside the existing ones; for the service water intake structure, they would construct an extension of an existing wall down to glacial till soil; and for the buried pipe, they would rebed some buried pipes and replace others.[34] But the fix for the problems with the cracking in the walls of the auxiliary building, which the NRC did not approve until 1982, required complex and costly tunneling, excavating, shifting the weight of the auxiliary building to temporary supports, and constructing a new support for the auxiliary building so that it did not rest on fill dirt.[35]

At the end of 1978, all of the details of the fixes for the fill dirt problem had not been determined, but it was certain that cost estimates would increase, completion dates would be pushed back, new public hearings would possibly be held providing new protest opportunities to the plant's opponents, and quality control would attract renewed attention. All of these challenges would mean that Consumers Power would need to borrow more money.

▪ ▪ ▪

Consumers Power's employees also saw a less tangible—but what they considered very real—effect from the sinking and cracking problem. Bob Wheeler, a Consumers Power engineer who worked on the Midland Plant, said, "After the sinking and cracking it became evident that the NRC wanted perfection in all of the construction and perfection can't be done."[36] It's impossible to know if the NRC ratcheted up its oversight of the Midland Plant because of the discovery in 1978, or if the greater oversight was the result of a more general change in nuclear regulation in the late 1970s. It is possible the NRC's enforcement of quality control at Midland may have had more to do with the particular history of the NRC rather than events in Michigan. The NRC started operations in 1975 and took over the licensing and rulemaking functions of the Atomic Energy Commission (AEC). But unlike the AEC, the NRC's statutory mandate was clearly focused on the safety of nuclear power and not on promoting its civilian use.

Soon after the NRC set up shop as the regulator of nuclear power, two things happened.[37] The first was a major fire at the Tennessee Valley Authority's Browns Ferry Nuclear Plant in March 1975, which raised the question of whether a single breakdown could initiate a chain of events that would incapacitate even the redundant safety features. The second event was the 1975 release of the Rasmussen Report, which concluded the risks from nuclear accidents were very small in comparison to other types of risks.[38] Critics faulted this report for failing to consider many things that could lead to a serious nuclear accident and for not supporting with data the optimistic conclusion in its executive summary. In January 1979 the NRC withdrew its full endorsement of the 1975 study's executive summary.[39]

But whether Consumers Power and Bechtel's poor record of quality control or the NRC's general program to closely monitor nuclear power caused increasing quality control oversight at the Midland Plant, Wheeler's observation that the NRC did continue to find fault with the quality of construction at the facility on numerous occasions after 1978 was certainly true.

Consumers Power's response to the continuing quality control problems in 1977 and 1978 was not to bring in new top management. Instead the executive management was concentrated in the hands of the existing leadership. John Selby said that in coming to Consumers Power as president in 1975 his goal was to run a company, and he achieved his goal in 1979. In April 1978, he was named chief executive officer when A. H. Aymond stepped out of that position.[40] Then in October 1979, Selby succeeded Aymond as chairman of the board, thus wearing three hats at Consumers Power: president, chief executive officer, and chairman of the board.[41]

The concentration of all Consumers Power's top positions in Selby between 1979 and 1984 not only meant that he was directing the utility's operations, it also meant that he was its critical public face. He was in charge of explaining the Midland Plant to the shareholders, the Michigan Public Service Commission, and politicians. These politicians would play a critical role in deciding Consumers Power's ability to recover its investment in the facility. In addition to the local opposition, financing difficulties, quality control issues, sinking structures, and public relations challenges, 1979 would present Selby with new problems to the completion of the Midland Plant that originated far from Midland.

CHAPTER 17

TMI2

Definitely Not Too Much Information

In 1968 Mary Sinclair opposed the Midland Plant because of her concerns that the proponents of nuclear power did not consider the safety risks of siting a nuclear plant close to where people lived and worked. But Dr. Wayne North and the Midland Nuclear Power Committee constantly stressed the *safety* of having nuclear reactors in the town's very backyard, saying the chances for a major nuclear accident were on the order of a billion to one.[1] Had Sinclair held a meeting of a committee in the fall of 1978 to brainstorm an event that would raise awareness of the safety risks of nuclear power, that plan could have included this: recruit major film studio to release thriller based on problems in the operation of a nuclear plant, then cast famous film stars to play the leading roles, and have the script emphasize the terrible effects of a nuclear meltdown by relating the effects to a large geographic area in the United States.

The meeting of the committee did not take place but if it had it could not have foreseen that in 1979 the movie actually would be produced and a real meltdown event would also occur, much less a snafu that happened in which the plant owners' and government officials' response to their real reactor problem began with a disingenuous release of information to the public, followed by a mistaken analysis of the problem, and throughout the entire incident release a muddled (and contradictory!) stream of information to the public.

The movie, *The China Syndrome*, was released on March 16, 1979, to over six hundred theaters.[2] It depicted a terrible problem and cover-up by utility officials at the fictional Ventana nuclear power plant. As discussed earlier, the term "China syndrome" describes a meltdown of a nuclear reactor core followed by the melt-through of the containment building. The hyperbolic term conveyed the image of a melted reactor core so hot and radioactive it would confirm children's belief that a hole in the ground could be dug all the way to China.

Despite its hyperbolic title, the film didn't depict impossible events and results from a nuclear accident. It is quite possible for a small problem at a nuclear plant to cause a cascade of events that are very serious: it is technically possible for the core of a nuclear reactor to melt down, and conceivable that an accident at a nuclear reactor can result in the release of radioactivity. The film's reception was undoubtedly enhanced because of the major film company that distributed it (Columbia Pictures) and two famous actors who played starring roles. The cast included Jane Fonda, winner of two Best Actress Academy Awards, who played the lead role of a television reporter; Jack Lemmon, also a two-time Oscar winner, who played the role of a plant supervisor who tries to make things right; and Michael Douglas, who played the role of a television cameraman. Both Fonda and Lemmon received Oscar nominations for their roles in *The China Syndrome*. Within two weeks of its release, Columbia reported that the movie had earned the highest income of any film it had ever released in a "non-holiday period."[3]

The film begins with TV reporter Kimberly Wells (Fonda) and her cameraman Richard Adams (Douglas) going to the Ventana Plant to get background material for a news series on energy in California. Part of their tour of the plant includes going to a visitors' gallery to observe the control room for the nuclear reactor. As soon as they arrive, they hear an alarm, and row upon row of lights start to flash. From their vantage point in the visitors' gallery, it is obvious that the men in the control room are highly concerned with and have no plan to control the problem.

The scene cuts to the control room, where we learn that the steam turbine has gone offline, which has caused the control rods to shut the reactor down. In spite of the reactor shutdown, it is essential that residual heat be removed from the reactor's containment vessel by cooling water to stop the core of the reactor from melting down. But the gauges on the control panel show the level of the cooling water in the nuclear reactor dropping. Supervisor Jack Godell (Lemmon) is forced to override standard protocols. After an extremely tense interval in the control room, the cooling water level starts to rise and the problem is dealt with.

Wells leaves the Ventana Plant and tries to understand what could have happened if the operators had not been successful in restoring cooling water to the reactor. She is introduced to physicist Dr. Elliot Lowell (Donald Hotton) who illustrates the effect of a melt-through of a reactor containment vessel by telling Wells that an accident there would render "an area the size of Pennsylvania" permanently uninhabitable.[4]

In Pennsylvania, just twelve days after the release of *The China Syndrome*, an alarm sounded in the control room of a real nuclear plant, Three Mile Island Unit Two reactor (TMI2). The problems on March 28, 1979, started with a failure in the non-nuclear part of the plant at 4 a.m. that caused pumps to stop sending water to the steam generators that remove heat from the reactor core. This failure caused the plant's turbine-generator and then the reactor itself to automatically shut down. In *The China Syndrome* it was only a few short minutes before the control room operators had a solution to the same problem. The real-life sequence of events at TMI2 took a lot longer and was much more complicated.[5]

After the initial alarm, the plant operators were unaware that cooling water for the TMI2 nuclear reactor was pouring out of a stuck-open valve. It was not until 6:22 a.m. that they were able to stop the loss of cooling water. In the absence of this water, the fuel rods in the reactor core ruptured and released radioactive fission products into the water and steam. At 6:56 a.m. the operators declared a site emergency, which meant that there was the possibility of an "uncontrolled release of radioactivity" within the plant's boundaries. At 7:05 a.m. Gary Miller, the Three Mile Island station manager, noting the alarmingly high radiation readings in the reactor containment building, declared a general emergency, defined as having the "potential for serious radiological consequences for the health and safety of the general public."

The media didn't get their first information on TMI2 from "official sources." Rather, "Captain Dave," a traffic reporter for Harrisburg's "Top 40" music station WKBO, picked up a notice on his citizens band radio at about 8 a.m. that police and firefighters were mobilizing in Middletown, three miles from Three Mile Island, and alerted news director Mike Pintek at his radio station. (Harrisburg is about thirteen miles northeast of the plant.) When Pintek called the Three Mile Island plant, the switchboard operator mistakenly put him through to the TMI2 control room, where an operator told him, "I can't talk now, we've got a problem." The person who answered in the control room told him to call the Metropolitan Edison Company headquarters (the utility operator and principal owner of TMI2).[6]

When Pintek called the headquarters, he was switched to Blaine Fabian, the company's manager of communications services. According to the news director, Fabian told him that the plant was shut down because of a feed water pump problem, the "general emergency" was a "red-tape" type of thing required by the Nuclear Regulatory Commission (NRC), and there was no danger to the general public. Pintek said he went with Fabian's story in his 8:25 a.m. broadcast on March 28th and tried to tone it down so people wouldn't be alarmed.[7]

In 1979 Americans were not all snuggled down in their own micro-news bubbles, Walter Cronkite was considered the nation's most credible journalist, and large numbers of Americans took what he said as fact. On Wednesday, March 28th, the night of the accident, Cronkite began the *CBS Evening News* with a report on Three Mile Island: "It was the first step in a nuclear nightmare." He continued, "A government official said that a breakdown in an atomic power plant in Pennsylvania is probably the worst nuclear reactor accident to date."[8] In the first three days of the accident, in addition to conflicting reports about the radioactive material that was being released into the atmosphere from radiated steam, the NRC generated a separate—and what turned out to be a fake—issue. Due to calculation errors, the commission had incorrectly concluded on March 29th that an 880 cubic foot hydrogen bubble that had formed in the containment vessel could contain enough oxygen to burn or explode and release radioactive material over the surrounding countryside.[9]

How did the Three Mile Island accident play in Midland? The front-page of the *Midland Daily News* in late March and early April included several headlines on the event. On Wednesday, March 28, the paper included "Steam Escapes at N Plant" heading an article citing Fabian, spokesman for Metropolitan Edison, who said a valve in the pressure steam system blew out. He stated that the accident automatically shut down the reactor, but some of the radiated steam vented into the atmosphere before the building was sealed. However, Karl Abraham, NRC public affairs officer, was reported to have said, "What they're seeing is extremely low radiation. It appears to be confined to inside the containment structure, the steel reinforced concrete dome around the reactor."[10] Was the radiation going into the atmosphere? Metropolitan Edison and the NRC had different stories.

The next day, Thursday, March 29th, the headline "Radiation Leak Cause Is Debated" alerted readers that radiation leaks from the Three Mile Island nuclear power plant still continued, as a debate grew over what was described as one of the most serious such incidents in the country's history. Officials of Metropolitan Edison said minor levels of radiation were measured that morning two to three miles away and that it might be Friday the 30th before leaks would be stopped altogether. "We concede that it's not just a little thing," Don Curry, Edison spokesman said, correcting the earlier statements that it was "just a little thing."[11]

Another headline on Thursday, "Accident Was Over in 8 Minutes," seemed reassuring. Within that short span of time, George Troffer, Metropolitan Edison manager of generation quality assurance, said the accident itself was over. But for hours afterward technicians deliberately released more radioactive steam into the atmosphere in a "feed-and bleed" emergency maneuver to cool down the stricken reactor. He said nothing escaped directly from the tightly-sealed reactor housing and that the high radioactivity inside it was not strong enough to penetrate the plant's walls. Troffer's statement contradicted reports from the NRC that radioactivity may have escaped from the reactor housing.[12]

The headline on Friday, "Will Leak Affect Midland Plant?," brought the event home. The article reported that Metropolitan Edison officials insisted that neither workers nor members of the public were harmed by Wednesday's radioactive release. But nuclear opponent Sinclair wasn't so sure. She pointed out that cancer caused by radiation doesn't show up instantaneously. She was also concerned that the incident happened at 4 a.m., yet the public wasn't told until 7 a.m. (actually as reported above the public was not told until 8:25 a.m.). "If you are a Pennsylvania mother and you knew that fetuses and small children were the most susceptible, you would have liked to have the option of getting out of the area. The real damage is the genetic damage that will not be traceable," she said.[13]

In the same article that reported Sinclair's comments, the *Midland Daily News* quoted two of Consumers Power Company's officials who said they thought the dangers of nuclear power were exaggerated in the minds of the American public. "People are extremely concerned about nuclear because it's a relatively new thing and an uncertainty," said spokesperson Norm Saari. The utility's corporate health physicist, Roger Sinderman said, "The reactor was isolated. There were no injuries to employees or the public. There can be an auto accident that kills 12 people that no one hears about, yet this nuclear plant that hasn't hurt anyone is all over the media. It's a good emotional issue because people are afraid of it."

Also on Friday, under the headline "Persons Near Plant Told to Stay Indoors," readers learned that Pennsylvania governor Dick Thornburgh urged residents within ten miles of the Three Mile Island nuclear power plant to stay indoors until further notice following a new release of radioactive gas from the crippled facility.[14]

On day 4 after the accident, readers learned from "N-Plant Bubble Reduced" that repeated venting of radioactive gases had reduced the size of a dangerous gas bubble at the top of Three Mile Island's crippled nuclear reactor, according to top utility executives. Governor Thornburgh had said at a news conference late Friday that no general evacuation of the area was necessary at that time, but he repeated his recommendation that young children and pregnant women move from within five miles of the plant. In spite of the fact that no formal evacuation order had been issued, many of the 131,000 people who lived within ten miles of the plant also packed their belongings into trucks, vans, and cars and drove away.[15]

On Monday, April 2, the headline "More Safeguards for Plant Here" reassured Midlanders that, although the Midland nuclear plant was "very similar" in design to the Three Mile Island reactor, which had also been designed by Babcock & Wilcox of Lynchburg, Virginia, according to Consumers Power officials, the Midland Plant would have additional safeguards by the time the reactor went into operation in 1981.[16]

The same day, another article reported "Bubble Shrinking; Temperature Drops." Metropolitan Edison spokesman Troffer told the Associated Press that as of 4 a.m. the bubble measured 210 cubic feet. At one point he said that the bubble was "gone," but he later acknowledged it still existed. Troffer made this claim even though on Sunday, John Harrington, another utility spokesman, said the company had been told by the White House "that all releases would be given out by the NRC." Also on Monday April 2nd, the *Midland Daily News* reported that on Saturday night in town, *The China Syndrome* had played to a full house at the Cinema I and II theaters on North Saginaw Road. The manager said the movie had been selling out regularly.[17]

By Tuesday, April 3, one week after the accident, things did not look good for the future of TMI2. "Reactor's Future Doubtful" read the *Midland Daily News* headline, followed by facts that preschool children and pregnant women were still being kept at least five miles away from the plant, an estimated fifty thousand people had left the area voluntarily, and businesses reported widespread absenteeism.[18] Finally on Wednesday, the headline "Catastrophe Threat Over" offered some relief. Governor Thornburgh had declared an end that day to the threat of "immediate catastrophe" in the nation's most serious nuclear accident. "We can heave a sigh of relief," he said on NBC's *Today* show.[19]

And what did the *Midland Daily News* editorial page—which had strongly endorsed the Midland Plant—have to say? When the immediate threat was over on April 4, on the editorial page, John Palen wrote,

> For as the questions pile up, one thing seems clear. The nuclear industry, perhaps overreacting to the extremism of some of its critics, has oversold the public on nuclear safety, and is now going to have to pay the price in a wave of skepticism and opposition. The public was told the chances of a major nuclear accident were remote. Remote or not, the dice rolled that way for people of the Three Mile Island area last week. The potential scale, if not the actuality, of nuclear disaster was brought home to the nation in startling immediacy.[20]

Of the many investigations of the Three Mile Island accident, the most prominent was the independent investigation announced by President Jimmy Carter on April 5, 1979. He appointed John G. Kemeny, president of Dartmouth College, as chairman of the President's Commission on the Accident at Three Mile Island.[21] The commission completed its report on October 30, 1979—which included a 179-page overview and nine volumes of task force reports that totaled more than 2,200 pages. The report identified a broad range of problems, some specific to Three Mile Island and others that applied to nuclear power in general.

The Kemeny Report found specific problems with the equipment installed at Three Mile Island, indicator lights deficiencies in the control room, quality control problems with the utility and the NRC, inadequate and shallow training of operators, and failures by Babcock & Wilcox and Metropolitan Edison to analyze safety problems and disseminate information on critical safety matters.[22] Particularly pertinent to the Midland Plant were the problems with Babcock & Wilcox's equipment that made it obvious that similar installations, like the one at the Midland Plant, would need to be changed. The Kemeny Report also said that the NRC lacked quality control procedures. This would haunt the Midland Plant project because after the report came out, the NRC would be motivated to be even more critical of everything in its purview than it had been in the past.

Consumers Power made a major public relations pitch to try to differentiate the Midland Plant from Three Mile Island. John Selby said,

> The Midland units were designed by Babcock and Wilcox, the same company that designed the two units on Three Mile Island, so there are certain similarities. Both are pressurized water plants, the megawatt rating is the same and the manufacturer

> of the major equipment is the same, but the plants aren't alike because the architectural engineers were different. We used Bechtel Power Company and Metropolitan Edison used Burns and Rowe, so many of the piping systems and auxiliary systems are somewhat different.[23]

The Consumers Power leader also recognized in April 1979 that the NRC hearings for the Midland Plant's operating license could prove difficult. He said, "In the operating license process, this accident gives our opponents a specific example of what can happen." Selby continued, "If I were an intervenor, I would try to make the most of this. Prior to this, they had to talk in supposition and had to assume certain things would happen in their arguments."[24]

Selby's efforts to minimize the changes to the Midland Plant's design as a result of the Three Mile Island accident were not successful. Although the opponents of the Midland Plant had no part in the making of *The China Syndrome* or in the accident in Pennsylvania, these events gave a major boost to the local antinuclear effort. Today it is generally believed that radiation from the Three Mile Island accident caused no injuries, but people feared the worst in 1979. The opponents of nuclear energy received tremendous support from what Cronkite said about Three Mile Island on the nightly news on March 30: "The world has never known a day quite like today. It faced the considerable uncertainties and dangers of the worst nuclear power plant accident of the atomic age. And the horror tonight is that it could get much worse. It is not an atomic explosion that is feared. The experts say that is impossible. But the specter was raised of perhaps the next most serious kind of nuclear catastrophe—a massive release of radioactivity."[25]

CHAPTER 18

People Power

Four Thousand March

In 1978 and 1979 Consumers Power Company was struggling to carry on the construction of the Midland Plant while dealing with the sinking and cracking problems and the reality that the Three Mile Island accident would result in requirements for major design changes. Although it was not until about 1973 that big national public interest organizations like the Sierra Club and Ralph Nader's Critical Mass Energy Project joined in the opposition to nuclear power, by 1978—prior to the Three Mile Island accident—these large national organizations along with many local groups were fully engaged in an effort to stop the construction of nuclear power plants.[1] The Union of Concerned Scientists (UCS), which included both scientists and engineers among its members, was formed in 1969 and contributed ongoing comments on nuclear power safety issues. For example, in 1975 the UCS sent an antinuclear statement to Congress with two thousand signatures.[2] In the meantime, what success were the opponents of the Midland Plant having in their efforts to shut down the construction of the Midland Plant?

At the height of its membership, the Saginaw Valley Nuclear Study Group that Mary Sinclair had helped organize in October 1970 consisted of only about thirty members.[3] But throughout the 1970s Sinclair persisted. She kept up a continuing schedule of speaking out against the Midland Plant, and by 1978–1979 the opposition to it had metastasized. During a cold rain in the spring of 1979, four thousand people marched down Main Street in Midland protesting the nuclear facility.

▪ ▪ ▪

Midland resident Sara Hughes had become interested in Sinclair's work when her daughter came home from high school and told her about a talk Sinclair had given at school that day. Hughes said she remembered thinking that if only half of the information Sinclair had shared was true, the twin nuclear power reactors planned for Midland presented unacceptable safety risks.[4]

Prior to her daughter's report, Hughes had become sensitized to the issue of nuclear radiation when a neighbor, who was a chemist at Dow Chemical Company, had shared with her his concerns about atmospheric testing of nuclear weapons. He said with all of the radiation from those tests, there could be a generation of children with leukemia. He also believed that a low-pressure weather system could keep any radiation leaks from the Midland Plant at ground level and expose Midlanders to radiation.

Years before, on March 1, 1954, the United States had tested a hydrogen bomb at Rongelap Atoll in the South Pacific, and by 1957–1958 the country had conducted a total of eighty-six known nuclear explosions in the air.[5] In 1972, a nineteen-year-old Marshall Islander, Lekoj Anjain, who had been an infant on Rongelap Atoll in 1954 when the hydrogen bomb test occurred, developed an acute case of leukemia. It was widely acknowledged that his radiation exposure from the airborne nuclear weapon test was the likely cause.[6] Hundreds of others also were likely the victims of the United States' airborne tests of nuclear weapons.[7]

The information from her daughter and her neighbor led Hughes to join the anti-nuclear forces in Midland and to pitch in to help Sinclair with typing and mailings. Helping Sinclair was not necessarily an activity in Midland in the 1970s that went without comment. Hughes recalled that when she took a batch of Saginaw Valley Nuclear Study Club envelopes to the post office for a mass mailing, the postal clerk noticed the return address and proceeded to give her a long, public dressing down in front of everyone about the fact that she was "standing in the path of progress."[8] Although Hughes remained active, she said she did develop reservations about how Sinclair was trying to influence public opinion. Hughes recalled telling Sinclair that she should make the things she wrote and said more accessible to a wider audience. Sinclair had responded that she was directing her efforts to the technical community.[9]

But not long after Hughes's suggestion, Sinclair did set in motion a new way to communicate the anti-Midland Plant message. In 1978 she gave a talk to a group gathered at the Guadalupe Center in Bay City. This talk struck a chord with a new audience, and a new path for opposition to Midland's nuclear facility opened up. Patrick Race, an activist in the anti-Vietnam War and civil rights movements, had invited five friends (Terry Miller, Patricia Murphy, Pam Rosengard, David Beckrow, and Edna Kay-Simons) to attend Sinclair's Guadalupe Center lecture.[10] Race and Miller said that before they heard Sinclair, they had both come to believe that nuclear power was bad, but her talk had impelled them to act.[11] So they formed a new organization, the Lone Tree Council.[12] The group choose its name to honor a version of a Native American legend described by Leslie Arndt in his book, *The Bay County Story*, about an aging, lone tree inhabited by a single white owl.

In the legend an ash tree, some eighteen inches in diameter, stood alone on the prairie about two miles south of Bay City. An owl was the supreme spirit that watched over and guided the destinies of the tribes of the Saginaw Valley from this lone tree, and as long as the tree stood and this angel kept its vigil, the tribes prospered. The legend also said when the spirit departed, the tribes would face destruction and disarray. For many years the owl continued its vigil, and the fertile valley provided ample resources for the people. But the tree died and the White man came to the valley and killed the owl. The Lone Tree Council considered the nuclear power advocates to be the new bad people who could bring destruction and disarray to the Saginaw Valley.[13]

Race and Miller both stated that although they had tremendous respect for Sinclair's writing, speaking, legal actions, and intervention in hearings, the Lone Tree Council would use other methods. "Direct action" was their way to keep nuclear power out of the Saginaw Valley—direct action meant demonstrations.[14] The council held its first demonstration on July 4, 1978, an event they called the March on Midland. By this time, the group had grown to sixty members. A picture of the demonstration published on page 1 of the *Midland Daily News* the next day shows about that number of people marching in an orderly line with a neatly lettered sign that said "STOP NUCLEAR POWER NOW."

Race said the marchers consisted of people he knew through his days of Vietnam War protests. He said that people with Viet Cong flags in the anti-war demonstrations had not contributed to the effort to win friends to the cause of stopping the United States' participation in the Vietnam War. Therefore, he felt it was important to keep control by vetting protesters so as not to have anyone who would get out of line and undermine the purpose of a protest, which was to attract people to the cause.

The anti-nuclear protests continued with a second demonstration near the gates of the Midland Plant on November 6, 1978. This time three hundred opponents of the plant participated. Sinclair was reported to have provided the emotional high of the demonstration in a speech questioning the safety and economy of nuclear power.[15] Although the founders of the Lone Tree Council organization believed that Sinclair liked to confine her efforts to writing, speaking, legal actions, and regulatory interventions, it seems evident that by 1978 she had warmed to the idea of reaching a wider audience with marches and demonstrations. Also mentioned in the description of this event was a young man named Michael Moore, who read a "Declaration of Nuclear Resistance," and then attached it to the gate of the site.[16]

Moore had set out at age fourteen to be a Catholic priest, a career aim he gave up after one year at St. Paul Seminary in Saginaw, Michigan. He went on to pursue a career of social activism and was elected in 1972 at age eighteen to his local school board. After a brief stint studying journalism at the University of Michigan–Flint, he went to work at the *Flint Voice*.[17] This newspaper's mission was to be an alternative outlet of informative, timely, compassionate articles about the city and the surrounding area that were not available in the local media. Although the *Flint Voice* claimed it had no editor or hierarchical structure, Moore functioned as the editor.[18] By 1978 and 1979 he was also active in an organization called the Huron Alliance, which summed up its views on nuclear power with the statement: "No material gain, real or imagined, is worth the assault on life itself that nuclear power represents."[19] Protests in Midland that followed the Lone Tree Council's event on July 4, 1978, were tailor-made for Moore and the Huron Alliance. He would go on to gain national notice with the 1989 documentary *Roger and Me*, followed by a whole series of films.

Race said he had been somewhat reluctant to join forces with Moore because he feared he would probably lose control, but the Lone Tree Council did eventually decide to join forces with the Huron Alliance to plan a demonstration for spring 1979. The council attended the alliance's meeting in November 1978 to plan an April 21 anti-nuclear protest in Midland. By this time, twenty-two anti-nuclear groups had indicated they wanted to join the demonstration. At the planning meeting the organizers anticipated that five hundred to eight hundred people would march.[20] In the last few weeks before the march, the number of sponsoring groups grew to forty-four, including the Public Interest Research Group in Michigan, the Great

FIGURE 13. March along Midland's Main Street protesting the Midland Plant, April 21, 1979. (Picture courtesy of Walter P. Reuther Library, Archives of Labor and Urban Affairs, Wayne State University)

Lakes Energy Alliance, the Flint branch of the American Civil Liberties Union, and United Auto Workers Local 599-Buick.[21] Prior to the event, twenty-four-year old Moore spoke at length to the press, declaring, "I am sick and tired of seeing Michigan being used for nuclear experiments. Midland epitomizes the destruction of our environment. To place in the midst of the chemical nightmare Dow has created there two nuclear plants is an abhorrent thought."[22]

On April 21, the marchers gathered at Emerson Park in Midland. As the crowd started marching away, a light spring rain began to fall, and by the time the protesters headed out of town to the Whiting Overlook Park, they were in a heavy downpour. Although some marchers had come prepared with rain gear, many walked along in shirtsleeves, getting soaked.

In spite of the weather, it quickly became obvious that many more than the originally forecasted number of people would participate in the march. According to the *Midland Daily News*, "There were hundreds of signs and banners carried by the marchers with such messages as 'I Won't Be Fooled by Conn-summers' and 'No Nukes is Good Nukes.'" The article described the chants of the marchers as varied and colorful and clean, like the signs. "Hell no, we won't glow" (a play on

FIGURE 14. Michael Moore speaking at anti-nuclear demonstration at Whiting Overlook Park in Midland, April 21, 1979. (© 1979 MLive Media Group. All rights reserved. Used with permission.)

the anti-Vietnam war protest "Hell no we won't go") was one of the more profane ones. Other chants included "2, 4, 6, 8—we don't want to radiate," "No more nukes," and "Nukes kill."[23]

The march proceeded down Main Street, over the Tittabawassee River on the Poseyville Bridge, and south on Poseyville Road to the Whiting Overlook Park (a distance of about three miles). This site provided a good view of both the Midland Nuclear Plant and Dow Chemical's production facilities.[24]

The hill at Whiting Overlook Park provided a perfect platform for a call and response session led by Moore's chant, "Shut it down! Shut it down!" The crowd yelled back, "No more Harrisburgs"—referring to the town near Three Mile Island. Moore yelled back, "This plant will never open."[25] The program on April 21 also included a number of speeches by nationally known activists, including the Rev. Wally Kasuboski, who worked as an investigator for the negligence lawsuit brought on behalf of Karen Silkwood, an Oklahoma plutonium processing plant worker killed in a car accident on her way to blow the whistle on safety concerns; Igal Roodenko, past president of the War Resisters League; and Bob Anton from *Akwesasne Notes*, a nationwide Native American newspaper.[26] While the speeches were in progress,

FIGURE 15. Protesters at anti-nuclear demonstration, April 21, 1979. (Picture courtesy of Walter P. Reuther Library, Archives of Labor and Urban Affairs, Wayne State University)

two thousand helium balloons were released bearing the message, "If this card had been radiation you would have been zapped by now."[27]

The *Midland Daily News* reported, "While most participants were from out of town, including places like Detroit, Traverse City, Lansing, and Grand Rapids, a number of Midlanders were among the marchers instead of along the sidewalks."[28] The local paper gave no estimate of the size of the crowd, but the *Detroit Free Press* said four thousand protestors marched 2½ miles through a steady rain, stretching out over a mile.[29] An April 22 article in the *Chicago Tribune* also estimated the crowd at four thousand.[30]

The large number of attendees can be attributed to two significant events that had happened in March 1979 that sparked an unprecedented outpouring of opposition to the Midland Plant: release of the *The China Syndrome* and the accident at Three Mile Island. As I read the account of the march in Midland on April 21 of that year, it struck me that it served as a crude measure of how these two events had changed public opinion about nuclear power. Without them, it seems likely that this march would have drawn only a few hundred people.

▪ ▪ ▪

Can the protest march in April 1979 be taken as evidence that the people of Midland did not want the nuclear facility built? Enter in 1980 a twenty-nine-year-old Midland County Clerk, Michael D. (Mike) Hayes, who decided to run for the Midland seat in the Michigan House of Representatives, 102nd District. His opponent was Sinclair. In a 2017 interview Hayes said he had been concerned by his early polling, which revealed Sinclair's name recognition was 98 percent and his was only 35 percent. He took comfort, however, from the fact that the same polling showed that though people knew who she was, their impressions were 9–1 negative.[31]

Both candidates ran spirited campaigns. Sinclair tried to present herself as more than a single issue candidate, but letters to the editor in the *Midland Daily News* and general comments always associated her with the Midland Nuclear Plant. In April 1980, prior to making her decision to run, she said that she wanted to know if the hostility that was evident when she first spoke out against the Midland nuclear plant remained.[32] When the votes were counted in November and Hayes won by a 2–1 margin, it certainly seemed that people in the Midland area did not want a legislator who opposed the nuclear facility.[33] But when Hayes went off to Lansing to claim his seat in the state legislature, he found that the general attitude toward nuclear power was driven by the negative response to the Three Mile Island accident. Nuclear plants were not political positives in the rest of the state. In addition, he found Consumers Power doing a terrible job of lobbying to pass the cost of the Midland Plant on to its customers.[34]

CHAPTER 19

The 1980s

Problems Intensify As the Money Drains

In 1980 Michael Hayes, Midland's new man in the state legislature, found Consumers Power Company not spending enough time and money to lobby for the Midland Plant in Lansing. The utility's lack of effort in the capitol may simply have been the result of the fact that it had too many other problems—keeping in place the contract to sell power to Dow Chemical Company, negotiating with the Nuclear Regulatory Commission (NRC) over exactly what it had to do about the sinking and cracking problems, and continuing to borrow enough money to keep on building the facility. Besides all that, the management also needed to keep the company's mom and pop shareholders happy.

A review of Consumers Power's progress in 1980 to build the Midland Plant makes it evident that the major problems the utility had encountered from 1967 to 1979 still existed. Rather than solving them and moving on, management allowed them to fester as the company doggedly continued in its efforts to finish the facility.

As noted earlier, Dow Chemical's opinion of Consumers Power back in 1976 had not been good. "Pretty damned close to blackmail" is what one Dow Chemical attorney, Leslie Nute, called the utility's efforts to keep Dow Chemical from canceling its contract for steam from the Midland Plant. Nute's notes from a September 1976 meeting with Consumers Power's legal department head James F. Alahee read,

"Alahee made a naked threat that if Dow's testimony [in upcoming hearings before the Atomic Safety and Licensing Board] was not supportive of Consumers, and that results in suspension or cancellation of the construction permit, then Consumers Power will file a suit for breach of the steam contract."[1]

The Dow Chemical–Consumers Power relationship did not improve in 1977. At the February ASLB hearing Alahee had been referring to, Joseph G. Temple, Dow USA director of marketing, repeated the blackmail charge. The former manager of Dow Chemical's Michigan Division, he testified that corporate officials agreed the plant no longer held any advantages for their firm and that Dow Chemical had no confidence Consumers Power would either finish the plant by 1982 or stay within the current $1.67 billion budget. Temple said, "The only reason Dow remains tied to the Midland Plant is a Consumers threat of a $600 million lawsuit if Dow abandons the project."[2] Dow USA President, Paul Oreffice, essentially repeated Temple's testimony at the same hearing. He said the threatened lawsuit had weighed heavily on his decision to stay with nuclear power. He went on to label Consumers Power's offer to let Dow Chemical walk away from the contract for $400 million as "extortion."[3]

The utility was able to settle these 1976–1978 differences with Dow Chemical with a new contract on June 27, 1978. Under this new contract Dow Chemical retained the ability to take four million pounds of steam per hour, but was obligated to take only one million, down from a previous minimum of two million, and had a right to buy its way out of the steam contract. In addition, Dow Chemical agreed to buy 174,000 kilowatts (KW) of electricity from Consumers Power in 1982, down from previous commitments to buy as much as 300,000 KW. For the first time, in this new contract, Consumers Power committed to have the Midland Plant in operation by the end of 1984. Previously the utility had contended that Dow Chemical was obligated to buy steam from the Midland Plant regardless of when the construction was complete.[4] But this resolution of the conflict between the two companies was not to last. By the 1980s, it became increasingly apparent to Dow Chemical that Consumers Power would be unable to live up to the terms of that 1978 contract and supply steam to Dow Chemical in 1984.

▪ ▪ ▪

Earlier I described an elaborate program of tunneling and excavation under the auxiliary building intended to fix the sinking and cracking problems. What I did not describe were the details of that auxiliary building work and the difficulty Consumers Power had in reaching agreement with the NRC on the needed corrective action. On December 6, 1979, the NRC issued an order calling a halt to attempts by the utility to correct the "soil problems" at Midland. Consumers Power responded with a request for hearings, which was granted in March 1980.[5] By requesting the hearings, Consumers Power was not bound by the stop-work order, but made it clear that its work would continue with NRC concurrence. The soils hearings started in July 1981 and included citizen interveners Barbara Stamiris of Freeland (a small town twelve miles downstream from Midland on the Tittabawassee River), Wendell Marshall of Mapleton (four miles downstream from Midland on the same river), and Mary Sinclair of Midland. Lone Tree Council members were also active in the hearings.[6]

The *Midland Daily News* reported that Barbara Stamiris was a former kindergarten and first-grade teacher in Saginaw Township.[7] The NRC lead attorney, William Paton, commenting on her involvement, said that intervention by citizens in nuclear licensing matters was "very, very healthy," and that intervenors tended to keep regulators' feet to the fire. However, the NRC's lawyer also noted that Stamiris's lack of legal training caused some difficulty inasmuch as her many questions drew objections from both Consumers Power and the NRC on grounds of improper form or content. He also noted that she had a "tremendous impediment" in that the NRC couldn't supply her with the free copies of the hearing transcript she needed to draft follow-up questions because the NRC's funding for procedural assistance to intervenors had dried up after President Ronald Reagan took office in 1981.[8]

The soils hearings provided a new forum for the critics of the Midland Plant to offer public testimony opposing its construction. Sinclair and founding members of the Lone Tree Council, Terry Miller and Patrick Race, took this opportunity to criticize the Midland Plant. Although she did not participate in all of the soil hearings, Sharon Warren, president of the Lone Tree Council, made a limited-appearance statement. She asked the board to consider the new issue of "ground collapse" and the size of the underground galleries that had been produced by Dow Chemical's brine mining in the vicinity of the Midland Plant. NRC attorney Ellen Brown told Warner that the NRC would respond to her concerns with testimony.[9] Two days later, Jeffrey Kimball, a seismologist with the NRC, presented evidence that the areas where Dow Chemical had used wells to dissolve salt and pump the brine to the surface would not be a problem, but that subsidence monitoring would be conducted over the life of the Midland Plant facility.[10]

▪ ▪ ▪

In addition to the public hearings, Consumers Power held private meetings with the NRC about the commission's requirements for fixing the problems with the auxiliary building. Rodney Wieland, a Consumers Power civil engineer, remembers a particularly long and contentious meeting involving the utility and the technical consultants retained by the NRC.[11] In charge of the meeting for Consumers Power was James Arthur (Jim) Mooney, a well-regarded, experienced engineer who held a master's degree in electrical engineering from Auburn University. Consumers Power had hired him away from the Alabama Power unit of the Southern Company, where he had worked for twenty years. While at Alabama Power, Mooney had been a construction manager during the construction of the Farley Nuclear Plant.[12] The meeting started in the morning and went on and on with much chatter and disagreement among the government experts. As the meeting dragged into the evening, Mooney—who had said very little and whose only act was to occasionally eat a mint from a packet in his pocket—finally said, "I've heard all about your models but I'm the project manager and I just want to get the job going again—what do I need to do?"[13]

The parties did not reach agreement on the fix for the underpinning project for the auxiliary building until 1982, and it was announced as an eighteen-month project. However, Mooney would tell the *Detroit Free Press* in April 1984 that this estimate was "not good."[14] He said a 1986 date was realistic for the completion of the underpinning project. And in fact, Consumers Power would say in April 1984 that the first unit of the Midland Plant would start operating in 1986, contingent on the underpinning being finished by early 1986.[15] The cost estimates kept rising, too. In 1983, the *Midland Daily News* reported that the underpinning project would cost $245 million, but at that time it was reported to be only 14 percent complete.[16] The following year, the *Detroit Free Press* published an estimate of $390 million.[17] I could find no estimate of the actual cost of the underpinning project, but with the completion date for the project pushed back from 1984 to 1986, it almost certainly would have cost more than the initial $245 million estimate provided in 1983.

■ ■ ■

The cost of the underpinning project was only one more contributor to the ever-increasing estimated cost to complete the Midland Plant. In March 1980 Consumers Power announced its commitment to finish the facility in spite of a new estimate that nearly doubled the earlier figure of $1.67 billion to $3.1 billion.[18] The latest number was blamed on changes required because of Three Mile Island: project engineering going from 85 percent complete to only 59 percent complete; miles of cable ripped out and rerouted, and five million feet of new cable needing to be added; and interest charges, insurance, and taxes during the extended construction period that would increase costs by $570 million.[19] By 1980 it became obvious that John Selby's earlier statements about the changes resulting from the Three Mile Island accident had been wishful thinking.

In the 1980s the need for more money made the Midland Plant a *Little Shop of Horrors* for Consumers Power as more money was always needed to feed the plant. The utility's management was forced to scramble to borrow the necessary funds. As of the end of 1980 the firm had invested $1.57 billion in the Midland Plant.[20] By the end of 1983, it had sunk over $3.43 billion in the project.[21] Yearly expenditures for 1981 through 1984 were $460 million, $660 million, $740 million, and $770 million.[22]

Not only were the sums of money needed to keep on building the Midland Plant huge, the interest charges Consumers Power was paying to borrow it were rising dramatically. Part of this rise was attributable to the rampant inflation of the late 1970s and early 1980s, which made the inflation of the Vietnam era look inconsequential (in 1980 the Consumer Price Index increased by 13.5 percent).[23] A related problem was that the utility was becoming an ever riskier party to loan money to. Consumers Power's financial struggles in the early 1980s were made more challenging by the general economic conditions and especially those in Michigan. A short recession in 1980 was followed by a deeper recession in 1981 and 1982. This period was called a W-shaped recession after the shape of the graph of the quarterly changes in GDP, which declined from a positive 3.166 percent in 1979 to a negative 0.257 percent in 1980 and then declined again from a positive 2.538 percent in 1981 to a negative 1.803 percent in 1982.[24] In September of 1982, Michigan led the nation with an unemployment rate of 14.5%.[25] The unemployment rate in Flint in December 1982 was 23.4 percent.[26] The state's dire economic conditions in the early 1980s were a far cry from the robust business climate that had existed when the Midland Plant was announced in December 1967.

Bruce Peck remembers one of the extremely depressing aspects of working on the Midland Plant in the early 1980s was being told that the daily interest costs alone were $1 million.[27] When I was told this number, it seemed to me that it must have been an exaggeration. But my examination of the interest rates Consumers Power was paying and the amounts of its investments in the Midland Plant by the early 1980s indicates that this figure was essentially correct. Having not anticipated this need for cash, as the decade progressed Consumers Power was having to come up with money to keep building the facility. In contrast, the other big utility in Michigan, Detroit Edison, was able to secure its financing to finish Fermi 2, which was under construction during the 1970s and 1980s.[28] In 1980 Consumers Power sold first mortgage bonds with up to 12.5 percent interest rate as compared to the 5.94 percent rate it had paid in 1967 when it announced the Midland Plant.[29] By December 1982 the company assumed that if it were able to issue first mortgage bonds, it would be at a rate of 14 percent.[30]

Not only were borrowing rates high in the 1980s, Consumers Power needed to run a gauntlet of requirements to borrow money. This gauntlet consisted of having earnings sufficient to meet ratios of earnings to debt proscribed by the company's articles of incorporation, gaining approval from the Michigan Public Service Commission (MPSC) to issue new debt, and maintaining high enough ratings from Wall Street to be able to sell new debt. As of December 31, 1981, the utility could not issue preferred stock because its earnings were not high enough to pass the tests contained in its articles of incorporation.[31] In the notes to the financial statements issued with its 1981 annual report, Consumers Power said its ability to issue first mortgage bonds (other than refunding issues) and preferred stock in the future would depend upon increased earnings.[32]

In February 1982, Consumers Power described its battle to gain approval from the MPSC to issue new long term debt. The company said that during the past several years it had experienced substantial delays in securing authorization from the MPSC to issue intermediate and long-term securities because the Michigan attorney general and others opposed to continuing construction of the Midland Nuclear Plant had intervened in the securities proceedings. However, in January of that year the Michigan Supreme Court issued an opinion holding that in securities proceedings the MPSC's inquiry is limited to whether a utility needs to issue securities to obtain funds for a lawful purpose. Because of this ruling, Consumers Power was able to gain favorable consideration of applications it had filed in December 1980 and January 1982 for authority to issue and sell securities.[33]

But the utility's money woes were not over. Although Consumers Power's earnings for the year 1983 allowed the company to sell first mortgage bonds, the annual report for 1983 noted that the downgrading of its securities by Standard & Poor's Corporation and Moody's Investors Service might limit its access to capital markets. Both ratings services had downgraded Consumers Power's securities in the fall of 1983.[34] Rating agencies' assessment of the finances of a corporation are extremely important since once a company's bonds have a junk rating, many pension and endowment funds are prohibited from owning them. Investment grades range from lowest risk to low risk to medium risk. Junk grade ratings range from high risk to highest risk to in default.[35] For Consumers Power's first mortgage bonds, Moody's downgrade was from Baa3 to Ba1 and Standard & Poor's from BB+ to BB (both of these new ratings carried the label "junk" and indicated high risks).

▪ ▪ ▪

Consumers Power discovered one innovative way to borrow money while it waited for approval from the MPSC. In 1982 it financed the Midland Plant through a sale and repurchase agreement with a special purpose corporation, Oakway IV, a wholly owned subsidiary of Connell Rice & Sugar Company.[36] Oakway IV secured a total of $165 million in credit from a consortium of ten banks, and Consumers Power conveyed assets to it in exchange for payments made to the company from the credit banks. Assets conveyed included the nuclear steam supply system on Unit 1, the Unit 1 turbine, the Unit 2 turbine, and various others not connected with the Midland Plant. Consumers Power retained the right to repurchase the nuclear steam supply system and the Unit 1 and Unit 2 turbines before the Midland Plant commenced commercial operation. During this year, Consumers Power paid $17.1 million in interest charges to the owners of Oakway IV.[37] Based on the available information, the utility paid its creditor interest charges equal to 10.4 percent ($165 million borrowed and interest for 1982 of $17.1 million). But relative to the amounts of money the utility needed to finish the Midland Plant and the finite amount of assets they had to pledge to future deals, Oakway-type financing was not the answer to solve all of the utility's need for cash.

In March 1982, in an attempt to obtain financing from sources outside the United States, Consumers Power created a subsidiary, Consumers Power Finance, N.V., organized in the Netherland Antilles. Its initial efforts appeared to have been unsuccessful, since the utility reported in the notes to its financial statement in the annual report that this subsidiary's only activities were those incidental to its organization, but press reports in 1983 noted that Consumers Power was successful with foreign financing.[38] In August 1983, Walter R. Boris, the company's executive vice president for finance, said that in 1982 he, along with President Selby and Treasurer Richard Griswold, did an exhausting tour of Europe, during which they visited nine cities in five days and met with 180 people representing about forty banks, "just to get them to notice." Boris stated that the trip was worth it since

Consumers Power was then able to borrow millions of dollars from European banks at less than the U.S. rate.[39]

In late 1983 Consumers Power finally received authorization from the MPSC to sell twelve million shares of common stock, six million shares of preference stock, and $450 million of long-term debt and did obtain new funds from the sale of stock and debt worth $621 million that year.[40] By December, Consumers Power said it could issue first mortgage bonds and assumed that they would carry a 15 percent interest rate.[41]

■ ■ ■

Even given the astonishing rise in the cost estimates for the Midland Plant and the increasing difficulties of continuing to finance its construction, Selby remained firm that the course Jim Campbell had set for Consumers Power to build nuclear plants in the 1960s was still the correct one in the 1980s. In March 1980, Selby said, “Considering the scarcity and the cost of oil, it should be reserved for transportation, leaving coal and uranium as the only realistic fuels we can consider. Natural gas is reserved for heating, and other sources of energy are not now sufficiently developed or economical to play a major role.” He also said, “Consumers eventually plans to get about half of its power from nuclear plants and half from coal.”[42] Implicit in the nuclear course was the assumption that Consumers Power had the political clout to gain MPSC's approval to pass the cost of the Midland Plant through to its customers.

It should be noted that Selby's explanations of the company's direction in the early 1980s were sufficient to keep one constituency happy: the loyal mom and pop shareholders who continued to purchase its stock. At the end of 1981, twenty-eight thousand shareholders were taking advantage of the Company's Dividend Reinvestment and Common Stock Purchase Plan in order to purchase additional shares of common stock—either by investing their dividends or by making cash payments. This number represented an increase of 8 percent over 1980. During 1981, plan participants reinvested $14,655,000 in dividends and made cash payments of $3,270,000 to purchase 1,012,480 shares of common stock.[43]

On January 6, 1982, Consumers Power's board of directors rewarded the dedication of its shareholders by increasing the dividend on common shares from fifty-nine to sixty-one cents per share.[44] Commenting on the future of the utility's sales following the notice of dividend increase, Selby said, “As we enter the first part of 1982, we expect sales to remain relatively flat, at least until mid-year. But Michigan has the ability and the resources to get moving again.”[45]At this moment

Selby needed to talk about "getting moving again" because, as he said in the first three sentences in his report to shareholders on February 12, 1982, "During 1981 the economy of our service area remained depressed. Auto production declined to its lowest level in 20 years. Unemployment, together with the cost of just about everything, went up."[46]

In October 1983 Consumers Power again increased the quarterly dividend on common stock, this time from sixty-one to sixty-three cents per share. In a February 3, 1984, letter accompanying the 1983 annual report, Selby said, "As our efforts to obtain timely rate relief and to complete the Midland Plant reach fruition, we expect to provide still further assurance of a reasonable return on shareholder investment."[47] Although the company's mom and pop investors seemed to stay convinced of its future, by April 1984 the price that new investors were willing to pay for stock did not reflect Selby's positive view. Its stock closed on April 9, 1984, at $9 per share, the price having fallen by more than 50 percent since October 1983, when Consumers Power had issued new common stock at more than $20 share.[48]

Quality Control Problems

Déjà Vu

Not only did the Consumers Power Company have contract problems with Dow Chemical Company, soil issues, and financing challenges as construction of the Midland Plant dragged into the 1980s, but it also was déjà vu for quality control and quality assurance problems. In March 1980 the Nuclear Regulatory Commission (NRC) found problems in heating, ventilating, and air conditioning work at the plant, violations that cost the utility $38,000 in fines.[1] Then in October and November 1982 an NRC inspection revealed an enormous backlog of quality control inspections and discrepancies between design drawings and the as-built conditions for the diesel generator building. The equipment inspected included structural steel used in the heating, ventilation, and air conditioning system intake fans and support system monorail; exhaust pipe, exhaust hangers, and exhaust silencers; diesel generator control panels; diesel generator air start piping and hangers; structural steel used as an attachment point for other equipment; and electrical raceways (cable trays) and wiring. This inspection cost Consumers Power $120,000 in fines.[2]

These fines for quality control violations did not go unnoticed by the media. On December 14, 1982, reporters Dan Rather and Robert Schakne broadcast an extensive story on the Midland Plant on the *CBS Evening News*.[3] Rather opened with, "About the Midland atomic power plant in Michigan, the Nuclear Regulatory Commission said a pattern of repeated, flagrant, and significant quality assurance violations [existed] at the Midland Plant. That was back in 1973; now almost a decade later the plant remains unfinished." Schakne said, "Cost overruns—ten times over budget, delays—eight years behind schedule and construction errors—affecting critical safety systems." He went on to report that the October 1982 NRC review of the Midland Plant discovered a near breakdown in inspections. To document the claim, a short interview with the NRC's James Keppler followed Schakne's report. Keppler said that fourteen thousand to fifteen thousand inspections had not been done for completed construction work.

The *CBS Evening News* story also featured a person behind a screen who said he had worked as an electrician at the Midland Plant and had seen trash, apple cores, and banana peels put in steam pipes and then not removed. The implication was that the lines would become plugged when the facility started up. Consumers Power and Bechtel Corporation refused to comment on the report, so the claims of the anonymous electrician went unrebutted. The report concluded that events at Midland suggested a case of industrial mismanagement by two major American corporations on a multibillion dollar scale.[4]

The NRC and the news media were not the only organizations that opined on the quality control at the Midland Plant. In 1981, Sharon Warren of the Lone Tree Council contacted the Washington, D.C.–based Government Accountability Project (GAP), which collected information from whistleblowers on government or public projects and then published reports on their findings.[5] In March and May of 1982 GAP representatives made trips to Michigan to investigate the Midland Plant, question witnesses, and collect affidavits.[6] The project released its report in June. Citing confidential affidavits from former workers, the report identified problems including improper welds, poorly trained inspectors, unqualified welders, debris in pipes, and noncompliance with blueprints, as well as alcohol and drug abuses.[7] Billie Garde, GAP's director, detailed these findings in testimony before the U.S. House of Representatives Subcommittee on Energy and the Environment of the Committee on Interior and Insular Affairs in June 1983.[8]

The author of one affidavit represented that he had worked for seventeen years in engineering, most recently at six nuclear plants, and that he held the title of the senior quality control engineer for nuclear welding. He stated that he had been a member of the American Society for Quality Control and had published several books on welding and structural steels.

> My reason for making this statement is to express my deep concern over the quality of construction at the Midland nuclear plant, where I was terminated in March of 1982 for persistently bringing defects in construction and specifications to the attention of my superiors.
>
> . . . My experience is in welding inspections. When I first came to Midland, in December 1981, I reviewed Bechtel's specifications and procedures. I was astonished to see that in numerous places, Bechtel had established standards which fell below the ASME [American Society of Mechanical Engineers] code . . .
>
> . . . High-pressure piping, which contains up to 1500 pounds per square inch, is very vulnerable material. It reacts like a balloon to a pinprick. A weakness in any part of the piping is a danger to the entire length. Because of this, I was very concerned to discover that many welds in the piping had been improperly ground down, grinding down the pipe wall thickness along with it.
>
> This was not only a violation in itself. It was part of a large problem having to do with inspections of the parent metal for the piping systems. In small bore piping, the only way to inspect the inside of the piping for corrosion is to take what is called a thickness and materials (TM) reading. This is a time-consuming process if done correctly. To the best of my knowledge, the Bechtel QC [quality control] inspectors rarely took the time necessary to do this type of verification. They usually relied on visual inspection only.
>
> Visual inspection can detect corrosion only on the outside of the piping . . .
>
> When I performed a thorough inspection myself of the piping, using TM readings for the inside of the pipe wall, I discovered extensive corrosion. Although the QC reports appear to assure that the piping is of safety-grade quality, these reports fail to reflect the problems of the piping systems which I discovered. To allow severely corroded piping to be approved for safety-related systems is in my opinion inexcusable, and certainly very dangerous to the successful operation of the plant. . . .
>
> Bechtel has a cost-plus contract, and had routinely wasted large amounts of money because they have little incentive to do the work right the first time. Each time further expenditures are required to redo work, it adds to their fee. I have seen work ripped out because of shoddy installations, redone, and then ripped out and redone again because it still wasn't right. One QC engineer, who has been

at Midland since the beginning, told me that over 90 percent of the piping in the entire plant has had to be cut out and replaced at one point or other . . .

My alarming experience with the field welding and the QC inspectors led me to speak to my boss, [redacted] in December and January . . .

On Friday, February 26, [redacted, a senior official] called me into his office and told me I was terminated. [Redacted] was also there, and the two of them told me that I hadn't been able to adjust to the way things were done at Midland, and so they would have to let me go.

A second affidavit came from a person who said he had graduated in 1969 with a bachelor of arts degree in anthropology from the University of Michigan and then in December of 1973 was indentured as an apprentice with International Brotherhood of Electrical Workers local 58.[9] He represented that he completed his apprenticeship in December 1976 and was currently a State Licensed Journeyman Electrician. In that capacity he was employed by the Bechtel Power Corporation in May 1978 at the Midland Nuclear Plant.

The "cost-plus-contract" that Bechtel has with Consumers makes it easier to understand their philosophy. The contract pays for time and material—the more material used and the longer it takes to install it the more money they make. As long as Bechtel can convince the customer, Consumers Power, that they are making progress and the delays and overruns are not their fault, then Bechtel can just continue to "ride the gravy train", as it were. The same philosophy permeates all the way down the ranks but with an added corollary: The less work done, the further behind you get. The further behind, the more likely you are to get overtime . . .

I recall one instance during which I worked two days without a job to do and was praised for my industriousness. My supervisor suggested that I "look busy" so I spent the two days measuring things—walls, ceilings, floors, cabinets. It didn't matter what I measured nor was there any need for the measurements. Everyone who saw me assumed I was working and that I knew what I was doing.

Another problem I think perhaps has not been recognized is the presence of debris in the small bore stainless steel lines. Some of these lines will carry contaminated coolant and waste products. I cannot begin to recount the times I have seen someone throw peanut shells, orange peels, banana peels or waxed paper into 2" and smaller pipes. Even the supervisors participated in this activity. Although these lines will be flushed in the final stages of construction, I find it

> difficult to believe that flushing these lines can blow the banana peels out the end, particularly because these lines are by no means straight nor even limited to a 360-degree bend . . .
>
> I was assigned to work with the quality control inspection sub-contractor, Comstock Engineering, in the fall of 1981. My job duties included the marking and identification of Q-conduit supports. I was to do the first inspection on the supports for safety-related electrical conduits . . . The Q-conduits supports were all required to be inspected by quality control because they all involved conduit necessary in an emergency shutdown of the plant . . . I found that many conduit supports had been in place and were supporting weight greater than permitted by the specifications. Seismic activity or even normal operating procedures could put stress on any overloaded conduit support. The resulting stress on the other supports could causse [*sic*] the entire conduit to collapse. The possibility of this happening at the Midland Nuclear power plant is very real because the conduits contain the five electrical systems that are needed in a safety shutdown. In an emergency, without the signal and power cables contained in the conduits I was working on the plant would just keep running . . .
>
> I finally decided to write a letter to the NRC regarding the quality control mistakes which I had been observing . . . I wrote this letter to the NRC in February, 1982 . . . I was terminated as an employee of the Bechtel Power Corporation at the Midland Nuclear Power Plant on March 31, 1982.

After GAP made its allegations public in June 1982, its report was criticized because the identities of the whistleblowers were not released. Did the two people who provided the affidavits have the technical expertise to opine on the items they observed, or an understanding of NRC regulations and how they constantly changed? Without knowing their names, there was no way to judge the validity of these documents. It was possible the affidavits were simply sour grapes of disgruntled employees who had been fired.

One letter to the editor published in the *Herald-Palladium* said, "What is GAP? I have been working in almost every phase of nuclear energy for the last ten years (submarines, reactor testing and utility operation), I have built, tested and operated reactors and have never heard of or dealt with GAP . . . my qualifications include a college degree in mechanical engineering and several years of engineering experience." The letter goes on to say: "Construction on the plant is about 80% complete, and should not be stopped due to unsubstantiated allegations orchestrated by a "think tank of the left."[10] Although the validity of all of GAP's claims about quality control and make-work problems at the Midland Plant were impossible to verify, in March 1984 Consumers Power received nine notices of minor violations from the NRC.[11] Many were the same as the ones contained in the GAP affidavits.

Then, on September 7, 1983, nine people who worked at the construction site were arrested on sixteen drug charges after a ten-week investigation.[12] In order to deal with the drug problem, Consumers Power took the initiative to invite an undercover, on-site law enforcement presence. An attractive female law enforcement officer was assigned to Bruce Peck's group of utility workers.[13] She was sent out with a clipboard on the work site and said she was taking a survey for Consumers Power. As she wandered the site, she had ample chance to talk to workers.

As the result of the undercover investigation, one Consumers Power employee, six people who worked for subcontractors, and two former workers were arrested. Charges included the delivery of LSD and cocaine and the manufacture and delivery of marijuana.[14] In a press report, the Midland police chief Donald Harcek said the investigation had been cut short when the undercover officer took a job with another police department. Police said that they had expected to find more drug problems if the investigation had continued.[15] Consumers Power notified the NRC of the drug arrests. Russ Marabito, public affairs officer for the NRC regional office in Chicago, said the commission considered the arrests "strictly a law enforcement matter." If there were nuclear fuel at the plant, the commission would automatically investigate, he said, but they wouldn't investigate at this point unless it found "reason to believe that (the case) might impact on safety."[16]

■ ■ ■

Not only did Consumers Power take the initiative to deal with the drug issues on the Midland Site in the 1980s, it also made two major efforts to address quality control problems. It announced the first of these programs, called the Quality Improvement Policy (QIP), in October 1981. In a statement published with this announcement, Consumers Power's site manager Don Miller said, "Doing the job right the first time, every time, is each individual's responsibility. The key to QIP is really an attitude change. We want all employees to think about the procedures which have been established and follow them carefully. Everyone should know the management is fully committed to the policy."[17]

Employees received awards for success in meeting the goals of QIP. In April 1982, John Strahl, Wally Sabo, and Bob Bull (who all worked for the Consumers Power

Construction Department) received QIP recognition for their efforts in completing the outage building on time and within budget.[18] This edifice was an office building (not a safety related structure) built at some distance from the other buildings on the site. This seems to have been the only instance of anything built between 1967 and 1984 for the Midland Plant that was completed on time and within budget.

The Consumers Power employees were not recognized for assuring quality—just for speed and frugality. Later events would demonstrate a basic problem with the outage building.[19] It flooded in September 1986, June 2017, and May 2020.[20] It was not located in the area where thirty-four feet of fill dirt had been added to create the island above the floodplain of the Tittabawassee River where the main structures of the Midland Plant were built. Watermarks on the first floor of the outage building show that in 1986 flood water was within two feet of the ceiling.[21] A *Midland Daily News* picture published online on June 30, 2017, also shows the same building underwater, while the area of the thirty-four feet of fill dirt addition is high and dry.

Consumers Power initiated a second effort to deal with construction standards at the Midland Plant in December 1982 in response to the deficiencies the NRC found in the emergency diesel generator building in November of that year. The utility reacted to these inspection findings by halting construction on most safety related systems, laying off 1,040 Bechtel workers, and commencing a complete cleaning of all structures. It also committed itself to what was termed a 100 percent over-inspection of safety-related work already installed at the plant.[22] Company officials called this effort the Construction Completion Program. The details included ceasing construction and withdrawing equipment from the auxiliary building, the diesel generator building, and service water building, followed by thoroughly cleaning each facility and then conducting a detailed inspection, followed by remobilizing for completion on a system-by-system basis.[23]

In spite of these efforts, in June 1983 the U.S. House of Representatives Subcommittee on Energy and the Environment of the Committee on Interior and Insular Affairs convened an oversight hearing to consider "Quality Assurance at the Midland Nuclear Powerplant." Chairman Morris K. Udall (Democrat from Arizona) presided over this hearing. He opened the session by saying, "Our witnesses today will provide information on the nature and scope of the quality assurance problem at the Midland site. I hope we'll be able to get an idea of the nature and significance of the problem and whether the NRC is on top of the situation."[24] Under close questioning in the hearing, Consumers Power's president and CEO John Selby testified that Bechtel was responsible for the soils problem:

MR. SEIBERLING [JOHN F. SEIBERLING, DEMOCRAT FROM OHIO]: Who was responsible for assuring that the building foundations and soil underneath them was properly done?

MR. SELBY: Well, the Bechtel Power Corp. was the contractor.

MR. SEIBERLING: Bechtel?

MR. SELBY: Bechtel. And United States Testing was the testing organization, not a part of Bechtel, that was doing the testing on the soil samples. And we would expect that Bechtel, with responsibility would also be responsible for assuring that the proper soils had been placed.

MR. SEIBERLING: Do you feel that they did their job properly?

MR. SELBY: Well, obviously when the end result doesn't turn out right, I can't feel the job was done properly.

MR. SEIBERLING: Well, certainly that's about the best test I know of: Results.[25]

Later, Selby was asked about Consumers Power's overreliance on Bechtel.

MR. SEIBERLING: Now in the summary and conclusions of overall effectiveness by ACRS [Advisory Committee on Reactor Safeguards] they say that region III [Nuclear Regulatory Commission Region III] inspections staff—I'm now quoting "Region III inspection staff believe problems have kept occurring in Midland for the following reasons. One, overreliance on the architect/engineers [Bechtel]." Would you care to comment on that?

MR. SELBY: Well, there may have been times when that was a consideration. We have become more and more involved, as time has gone on. As we have seen some of the things happen, as we developed our own capability in order to go in and take a more significant role. I don't believe that's the case today.[26]

James Keppler, Regional Administrator of the NRC region III Chicago office, also testified at these hearings. He was the NRC official primarily responsible for monitoring quality control at the Midland Plant. He began his testimony by saying that he had previously testified before the NRC's Atomic Safety and Licensing Board in 1981, and at that time he had noted that many significant quality assurance problems had occurred in connection with Midland but that Consumers Power and its contractors had taken corrective actions. It was his opinion then that the problems did not indicate a breakdown in the implementation of the overall quality assurance program. Keppler concluded, "The thrust of my testimony at that time was that I had confidence in the Consumers Power Co.'s QA program both for the remedial soils work and the remainder of the construction."[27]

But now Keppler pivoted to a strong criticism of the utility. In April 1982 he had been made aware that additional significant quality assurance problems had been encountered, and as a result he had notified the Atomic Safety and Licensing Board that his 1981 testimony would have to be modified; that he had created a special section within the region III office solely to handle the Midland Project; that he had informed Consumers Power that additional measures were required to assure the quality of the plant; and, finally, that he had conducted a comprehensive inspection of systems and components in the diesel generator building. This inspection ultimately led to the major "stop work" action in December 1982.[28]

Summarizing his plans for future reviews of the Midland Plant, Keppler said, "Where we stand today, Mr. Chairman, is that Consumers Power Co. has proposed a number of changes which the staff is reviewing, that will consist of a backward look at the completed construction to date; will consist of a program to complete the plant and complete any necessary rework that may be done—all of this overviewed by a third-party organization in addition to the NRC."[29] On a less than optimistic note, he continued, "We believe these programs, when we complete our review of them and approval of them—we hope that these will provide confidence that the project will be completed satisfactorily. In any event, we want to assure this committee that the NRC will not issue a license for this facility until we are satisfied the construction has been completed properly."[30]

The NRC did add a third-party review of the Midland Plant, and in April 1984 Keppler told a five-member NRC panel in a special briefing that this examination of Consumers Power's work restored his confidence in the utility. This Construction Completion Program included supervision by Stone & Webster, a Boston-based engineering firm.[31] The *Lansing State Journal*'s headline in the story that reported Keppler's statement read, "Consumers Finds Ally in NRC."[32]

The statement in 1984 that the NRC now had confidence in the quality of Consumers Power's construction efforts at the Midland Plant was nothing new. As detailed earlier, the utility had problems pouring cement in 1970, cadwelding in 1973, stringing tendons in 1977, compacting soil in 1976–1977, constructing heating/ventilation/air conditioning systems in 1980, and following construction diagrams for the emergency generator building in 1981. Over and over, the Atomic Energy Commission (through 1974) and then the NRC (beginning in 1975) found problems with both Consumers Power and Bechtel's construction practices and then allowed them to continue building the plant.

Sharp criticism of Consumers Power and the Midland Plant continued in 1984. In the same April 1984 briefing when Keppler told the NRC panel he had confidence in Consumers Power's ability to finish it, GAP director Billie Garde pointed out that in the past after Keppler made similar statements, the company had kept making mistakes. She concluded, "Consumers Power Company, its Midland project and its regulatory history should be an embarrassment to this commission. It is an engineering nightmare, a financial disaster, and a construction boondoggle. It is a political headache and a regulatory muddle."[33]

CHAPTER 21

Brink of Bankruptcy

No Sales Contract, No Cost Recovery, and No Money

By October 1983, Consumers Power Company acknowledged the toll the Midland project had taken on its finances. The company's chief financial officer (his official title was executive vice president, finance and corporate affairs), Walter Boris, said, "Our problem is a cash problem, compounded by the tremendous expenditures we still have ahead of us." He added, "But I don't view this company as being on the brink of bankruptcy," saying, "only marginal investors have been chased away."[1] Consumers Power's chairman of the board, president, and chief executive officer John Selby also said in that month, "We are still generating enough cash . . . so we can pay our common stock dividend. Our situation with regard to the balance sheet doesn't in any way represent a desirable case . . . [but] I think we have the wherewithal to complete that plant."[2]

Boris and Selby made these optimistic statements without regard to two major problems. The first problem was that Dow Chemical Company had given notice in July 1983 that it was terminating its contract with the utility. This not only had financial implications for Consumers Power, but it also deprived the utility of important public support from Dow Chemical and the people in Midland. The second problem was Consumers Power's implicit assumption that the Michigan

Public Service Commission (MPSC) would allow it to recover most of the Midland Plant costs from its customers.

In addition, Dow Chemical sued Consumers Power for $60 million in damages, charging that the utility had misled Dow Chemical about construction delays and cost overruns. In describing the reason for the contract termination, Hunter W. Henry, president of Dow Chemical U.S.A., said, "Spiraling cost and interminable delays have made the cost of steam from the nuclear facility intolerable. Its use would make our Michigan Division non-competitive with other locations and competitors, and seriously jeopardize the continuation of major manufacturing operations here."[3]

Then in November 1983, Consumers Power filed an electric rate case with the MPSC asking to be allowed to add $490.7 million (27.7 percent) to the rates it charged its customers when the Midland Plant went into operation. The attorney general, the Michigan Citizens Lobby, and the Association of Businesses Advocating Tariff Equity (ABATE) all intervened in the case.[4] This meant that at a minimum the case would be dragged out while all of the opponents presented their evidence. As Boris had pointed out, Consumers Power had a cash problem, and regardless of the eventual decision in the rate case, the company needed certainty and needed it quickly so it could borrow more money to continue building the Midland Plant.

As Consumers Power entered negotiations with the MPSC in 1983, the company's attitude seemed to revert to A. H. Aymond's belief that the State of Michigan had to allow the company to increase rates in order to continue its expansion.[5] Selby's slightly more diplomatic response to questions about who should pay for the soil problems at the Midland Plant was, "Mistakes were honest, so our customers should pay. Electric customers should pay for soil problems because they were the result of honest mistakes and not fraud." Selby went on to say, "Unless it's the kind of thing you can tie to being fraud or clearly a top management kind of oversight or mistake, I think it's part of the job."[6]

■ ■ ■

Consumers Power faced a powerful and skilled adversary in its 1983 efforts to raise rates: Frank Kelley, then Michigan's attorney general, who made a practice of intervening in rate cases before the MPSC. The increase Consumers Power needed to cover the cost of the Midland Plant gave him an extraordinary opportunity to protect the captive utility customers from paying sharply higher electric rates. Kelley had been appointed in 1961 by then Michigan Governor, Democrat John Swainson, to fill a vacancy left when Paul L. Adams became a justice of the Michigan Supreme Court. Kelley was then elected in his own right as a Democrat ten times before his retirement in 1999. After retiring, Kelley said,

> Michigan's attorney general has the constitutional right and duty to independently sue on behalf of the public. As attorney general, I could also intervene in any case, anywhere in the state, in which I determined that the greater public interest was involved. This has been true ever since Michigan was first admitted to the union in 1837. That's not the case in every state; in some, attorneys general have far more limited powers. But I had the ability to use my office not only as a bully pulpit but to battle on behalf of the people.[7]

Kelley viewed his time in office as a period when he transformed the attorney general's role from a passive defender of lawsuits against the state to an active and muscular crusader for the rights of citizens. He listed some of his achievements:

1. instituting the first department of consumer protection
2. taking action that helped lead to the one-person, one-vote decision [Kelley issued a series of opinions, starting in 1962, that said the United States Constitution's requirement that each citizen has equal protection under the law meant one-person-one vote. The United States Supreme Court later in the 1960s said the same thing in *Reynolds v. Sims* when it established that legislatures must be apportioned according to the one-person, one-vote standard.]
3. helping initiate the Tobacco Master Settlement Agreement, concluded in November 1990 (this agreement mandated that nearly $6 billion would be paid to the state of Michigan over the following quarter century)
4. taking on and taming Michigan's public utilities (which in his view often acted as though they were something of a law unto themselves)[8]

Early in his career Kelley discovered that Michigan utilities were a good target for his consumer advocate activities. Rodney Boulanger, who worked as a rate analyst for a Michigan utility as a young man in the 1960s and methodically developed the numbers to justify its rates, remembers how, whenever a utility filed a request for a rate increase with the MPSC, Kelley would intervene. His first move would be to issue a press release complaining that the utility was seeking an obscene amount. The

Detroit papers would then publish an almost verbatim account of his press release. Then he would appear on television in Detroit. Once on camera, halfway through Kelley's harangue about the utility's rates, the TV interviewer would pick up the issue of the newspaper that had printed the press release and announce that what he was saying must be true because the interviewer had read it in the newspaper.[9] In his 2015 book Kelley wrote that his deputy, Leon Cohan, years after he worked for Kelley, admitted that "[Cohan] sometimes would actually even write stories for certain members of the press, some of whom were perhaps better at gin rummy than at understanding and explaining the fine points of legal actions. We never, ever, lied to reporters, and we tried whenever we could to make their jobs easier."[10]

Kelley had his critics. There were those who called him the "Eternal General" (he would eventually establish the record of being the longest serving state attorney general in the United States). Many of his foes were Republicans who were forever frustrated that he could get elected time after time when other Democratic candidates were defeated. His critics also did not appreciate his "consumer's advocate" efforts when he would attack utilities and businesses every chance he had. Some of his critics described his actions as like those of a heartless driver who, on a dark Michigan night, sees the driver ahead of him hit an animal. As the animal writhes in pain on the shoulder, the heartless driver sees the wounded animal as an easy target and then takes to the shoulder and runs it over again. Consumers Power was the writhing animal as it reached the point in 1983 when it desperately needed approval from the MPSC to recover its costs on the Midland Plant. One could possibly say, though, that Kelley thought if he could only get Consumers Power to stop building the nuclear facility, he could put the company out of its misery: a decision it had been unable to make on its own.

In 1983, the assistant attorney general, Hugh B. Anderson, told the press,

> Kelley began telling Consumers Power and Detroit Edison 10 years ago that their forecasts for growth in electric demand were "greatly overstated" and their plans to build more generating plants too ambitious. In 1974, Consumers was predicting electric load growth at 7 percent annually, the historic rate for many years. Kelley, in the same year, forecast 2 percent growth. And even he was proved optimistic—Consumers' annual load growth was 0.7 percent per year from 1973–82.[11]

Anderson also liked to launch fundamental criticisms of the utility. For example, also in 1983 he said,

> The name of the game in the utility industry has always been to build as much as you can and increase the rate base, since you are allowed a fixed return on the rate base . . . Consumers Power officials failed to realize that the rate increases they sought in the 1970s would dry up electric demand and make the Midland plant unnecessary. It's like none of them ever went to the first semester in college and heard about the elasticity of demand related to price.

He noted that Consumers Power's rates for its large industrial customers already were 30 percent above the Midwest average.[12]

In a letter to the editor in the *Midland Daily News*, John C. Carlson expressed his views on Kelley's continuing attacks on Consumers Power :

> Your statements really amount to grandstanding but they have had a devastating effect on the company, its stock and bond holders who are also customers, and finally on just the customer who does not own stock but does pay the bills. Partly because of your thoughtless statements, credit markets, the lifeline of all utilities, have virtually been closed to the company. Bond issues, preferred stock and common stocks have all been drastically downgraded by rating companies . . . I firmly believe that ratepayers should not be seduced by you into believing "oh boy" that Frank Kelley is really doing a job for us—he's really horse-whipping Consumers Power. In truth, in the long run you are doing a "job" on them. When you say "not one dime of the cost of the Midland [plant] should be passed onto the rate-payer"—you are grandstanding. Such machinations are designed to champion the "cause" of Frank Kelley—not the "cause" of the ratepayer.[13]

Anderson responded to the type of charges leveled by Carlson and others by saying, "If GM [General Motors] builds a plant to build gas-guzzling cars and people want fuel-efficient cars and won't buy them, they have to shut the plant down. That's the system. It may be unfortunate, but I wouldn't call it unfair. That's what investors get paid for, is taking those risks. If they don't want to take risks they should buy government treasury bonds at 8 percent interest."[14] In August 1983 Anderson also advanced the argument that the MPSC actually protected the utilities it regulated, rather than their customers, in granting utilities higher rates. He cited three specific examples when Detroit Edison and Consumers Power had been able to recover costs spent on cancelled nuclear plants. He claimed that the MPSC commissioners "voted to stick it to ratepayers" in allowing Detroit Edison to recover $88 million for its Greenwood nuclear plant (abandoned in 1980) and $13 million for its Fermi 3 unit (cancelled in 1973), and allowing Consumers Power to recover $20 million for its Quanicassee nuclear facility (closed in 1974).[15]

▪ ▪ ▪

The search for a political solution for the Midland Plant began in early 1984, when representatives from Kelley's office; the influential Michigan Citizens Lobby; the staff of the MPSC, led by a former Consumers Power executive, Roger Fischer; and ABATE began a series of secret meetings to discuss the facility and Consumers Power's request for an electric rate increase. By late March this coalition had reached an agreement that Consumers Power should cancel the Midland Plant in exchange for a modest rate increase and a financial survival plan.[16]

At this point in April 1984, the underpinning program for the auxiliary building was announced to be only one-third complete, even though Consumers Power had estimated in 1982 that it would be an eighteen-month job.[17] Now with this new estimate, the revised date for the completion moved back again, to December 1986, with the caveat that 1986 would work only if the underpinning job was indeed finished.[18] In April 1984 the estimated cost for completion had reached $6 billion.[19] Once again, the estimated completion date had been pushed further into the future and the estimated total price had joined a long history of ever-increasing estimates (see table 4).

In spite of Selby's and Boris's optimistic October 1983 statements about Consumers Power's financial condition, by 1984 the Midland Plant was bleeding the utility to death. As reported earlier, by the end of 1983, Consumers Power had invested $3.43 billion in the plant. The press reported on June 12, 1984, that the facility was costing Consumers Power $50 million per month and that it would run out of cash by mid-July unless it reached an agreement with the state.[20] On May 1, 1984, Chase Manhattan Bank vice president George Wick appeared before a meeting that included Selby, Kelley, Anderson, representatives of the MPSC, and representatives from a coalition of Consumers Power's industrial customers and a ratepayers' group.[21] Wick's message to the group was simple: he had failed in his effort to line up a $500 million line of credit for Consumers Power.[22] Speaking for the attorney general's office, Anderson said, "The banks weren't going to loan another dime to Consumers, except perhaps for some seasonal inventories [natural gas and coal purchases for the coming winter], until this whole thing is settled."[23]

In a reversal of his statements of October 1983, Selby admitted after this meeting

TABLE 4. Estimated cost to complete the Midland Plant made by Consumers Power by year.

YEAR	MONTH	ESTIMATED COST TO COMPLETE
1967		$267 million[*]
1969		$375 million[†]
1971		$500 million[‡]
1974		$940 million[§]
1975		$1.40 billion[‖]
1977		$1.67 billion[#]
1980		$3.10 billion[**]
1982		$3.39 billion[††]
1983	April	$4.43 billion[‡‡]
1984	April	$6 billion[§§]

In July 1984 Consumers Power made a final estimate of $5.6 billion to complete one reactor and leave one unfinished. Bernie Shellum and David Everett, "Consumers Cancels $5.6 billion Project," *Detroit Free Press*, July 17, 1984, 1A.

* "Consumers to Build $267 Million Nuclear Plant for Dow Power Use," *Midland Daily News*, December 14, 1967, 1.
† Betty Herring, "Nuclear Plant Called 'Greatest Investment in the State,'" *Midland Daily News*, June 27, 1969, 1.
‡ Jerry M. Flint, "Midland Mich. Wants Nuclear Power," *New York Times*, December 3, 1971, 24.
§ "Costs Are Soaring For Power Plants," *Holland (MI) Evening Standard*, January 9, 1974, 14.
|| "Midland N-Plant Cost Jumps 48%," *Midland Daily News*, April 12, 1975, 1.
\# Sandra L. Dickey, "Potential Problems at N-Plant Identified," *Midland Daily News*, January 19, 1977, 1.
** Barry Rohn, "The High Cost of Nuclear Power," *Detroit Free Press*, May 4, 1980, F1.
†† Consumers Power Company, *1981 Annual Report* (Jackson, MI: Consumers Power Company, February 12, 1982), 2.
‡‡ Robert H. Bork Jr., "Nuclear Plant Hits Another Delay," *Detroit Free Press*, April 13, 1983, 37.
§§ "N-Plant Bill May Go up by $1.5 Billion," *Midland Daily News*, April 10, 1984, 1.

that unless some agreement was reached on Consumers Power's recovering costs, the company would need to begin to shut the Midland Plant down.[24] That month Consumers Power made a proposal to the MPSC to complete one of the reactors, charge customers $3.5 billion of the cost, and be allowed to recover an additional $800 million for the abandoned reactor. In response to that offer, on June 11, the staff of the MPSC countered with a proposal that would have allowed the utility to complete part of the Midland Plant and recover $2 billion of its investment. Consumers Power rejected the offer.[25]

■ ■ ■

In the midst of all of these last-ditch negotiations, what was happening on the actual site of the Midland Plant were preparations for a "turbine roll." This operation happened on June 14, 1984, and consisted of taking steam from gas-fired boilers on the site of the facility and putting it through one of the steam turbines to produce electricity. The turbine was brought up to its full speed of 1,800 rpm for about fifteen minutes. Bob McCue, testing superintendent for Consumers Power, said, "The control room was just ecstatic this morning for 15 or 20 minutes."[26] This did produce the plant's first electricity, but since steam from gas-fired boilers was used to turn the turbine and generate electricity, this turbine roll had nothing to do with how the plant was designed to operate using steam generated by heat from the nuclear reactors. "No matter what's going on in the world, progress on the turbine roll has been very good," Consumers Power's Norm Saari said, continuing, "There continues to be an overriding concern about the future of the plant. But there are sites [*sic*] and sounds of a power plant in operation that is exciting to a lot of people."[27] However, this public relations stunt seemed to have no effect on the negotiations to save the Midland Plant.

On the day of the turbine roll, Michigan's Democratic governor James Blanchard said he had no intention of entering the negotiations over the Midland Plant. "The issues don't change by me sitting at the negotiations," he said. "The technical merits don't change."[28] But up-and-coming Republican politician John Engler jumped into the fray in late June to try to find a solution. Engler, who was at the time the Michigan Senate majority leader, met with Selby of Consumers Power and Fischer of the MPSC and asked them to make one final effort to come up with a plan to solve the Midland Plant problem. This was a smart political move by Engler, who would later be elected governor. Although he would ultimately be unsuccessful in his efforts to save the Midland Plant, he could say he had tried to do something to save the situation, which was more than Governor Blanchard had done.[29]

So, unwilling to accept the offer to let it recover $2 billion of its Midland Plant costs and unable to borrow any more money to continue construction, in July 1984, Consumers Power was out of options with no deal and no money.

1985

What Happened Next?

At 6:20 p.m. on July 16, 1984, John Selby announced from the company's Jackson office that Consumers Power Company was cancelling the Midland Plant project. The unanimous decision by the board of directors came after key industrial customers had rejected the utility's latest financial plan to complete one reactor and leave the other unfinished. The estimated cost of that final plan was $5.63 billion.[1] Consumers Power had taken seventeen years to convert a plant with an original estimated cost of $267 million into one with a final estimated cost of $5.63 billion. Jim Mooney, the laconic Consumers Power's engineer from Alabama, always gave the same five word answer when asked why the effort to build the Midland Plant failed: "Consumers ran out of money."[2] He memorialized his judgment in a paper he coauthored in 1988: "Consumers Power Company halted construction in 1984 due to their inability to continue to finance the project."[3]

The political reaction to the cancellation was swift and predictable. Midland's Republican state representative, Mike Hayes (who had won his seat in 1980 by defeating Mary Sinclair) said, "Some awfully short-sighted people seem to be winning out for their own political gain. If those people feel they've gained a victory, then they should be willing to accept the consequences for the 6,000 human beings who will be without work, and for the lack of reasonably priced power."[4] In addition to employment effects, the cancellation potentially held major consequences in other areas, one of which was tax revenues. "We've got double-digit unemployment here, and this will just make it worse," said Midland mayor Joseph Mann. "The real question for this city is whether Consumers Power will be able to pay the $2.7 million in property taxes on the plant, due September 15. That is a very large chunk of our operating budget of $14 million, and it would require a considerable readjustment in services if we don't get it."[5] Another effect was that within a few months the real estate market in Midland crashed. From 926 active for sale listings in the county in September 1983, the number rose to 1370 in September 1984. The average property sold for $59,947 in 1983, but dropped by more than $5,000 in 1984.[6]

One individual tradesman—representative of thousands—affected was Mike Gronski, who got his start at the Midland Plant. He asks everyone to call him "Mike the plumber." In 2016 recollecting his time at the Midland Plant, he said he almost starved to death working for the half pay of a plumber's apprentice.[7] He persevered through the apprentice program, though, and went on to do the most intricate work of his whole career. Gronski did not speak of his work on the Midland Plant as something that had happened in the hazy past; he spoke with great pride of the work he had done there and with intense bitterness toward the people who opposed the plant. In his view the opponents of the facility stood in the path of progress. His bitterness was still raw in 2016; it was as if the Midland Plant controversy was currently happening.

In defending his support for cancelling construction of the Midland Plant, attorney general Frank Kelley said:

> This action has outraged two groups—the utility's stockholders and employees of the Bechtel Corp., who are building the plant. Apparently the stockholders feel, despite all the evidence to the contrary, that this plant should be built. Rather than blame Wall Street for its decision to stop loaning their company money they find it easier to blame me for not agreeing to Consumers Power's request that I guarantee its future profits so that the loans will be granted. As much as I would like to be helpful, I simply cannot ignore my duty to the people of this state and agree on their behalf that Consumers' customers will pay for this excessively costly power plant.[8]

▪ ▪ ▪

Consumers Power's stockholders were indeed big losers with the cancellation of the Midland Plant. The price of the stock was pushed to all-time lows, and the dividend on the company's common stock was stopped in October 1984.[9] The price of the common stock, which had reached a high of $21.875 per share in 1983, plunged to $4.125 in the fourth quarter of 1984.[10] So much for its being a mom-and-pop stock to own for retirement.

Bechtel, however, did not miss a beat due to its drastically over-budget and behind schedule involvement in the Midland Plant. The engineering firm continued working on modifying and completing jobs in the nuclear business in the United States as well as working on nuclear projects in Korea and Japan.[11] In addition, Bechtel continued to manage all sorts of other huge projects, including extinguishing fires in Kuwait after Desert Storm, helping with reconstruction in Kuwait, performing a major role in Boston's Big Dig Project, and constructing a highway through Albania. It did not seem to matter that it had been the company that oversaw the construction of a facility that ended up with an estimated final cost of $5.6 billion dollars (twenty-one times the original estimate).[12]

In March 1985 Consumers Power announced Selby's retirement, effective November 1. At the time one financial analyst, David Rudakas with Duff & Phelps Inc. Rating Co., delicately described the situation: "With these kinds of problems to the point where a regulated utility is forced to forgo common [stock] dividends and has a distinct possibility of bankruptcy, I don't know if I'd go so far as to say they start to look for management changes, but management changes certainly come to mind."[13] Also in March, Walter R. Boris, Consumers Power executive vice president of finance and corporate affairs, announced he would take early retirement. Following Selby's resignation, the Michigan Public Service Commission (MPSC) urged the utility to bring in new management "to chart the future course" of the company. "Boris's resignation gives the utility another chance to do that," MPSC commissioner Matthew E. McLogan said. "We hope and fully expect the company will take advantage of their opportunity."[14]

In September, Consumers Power reported that Selby had agreed to continue as a consultant. Jeri Kornegay wrote in the *Jackson Citizen Patriot*, "He will particularly assist the utility in defending itself against the Dow Chemical lawsuit on the cancelled Midland nuclear plant contract, a suit by a group of stockholders, and against charges that attempting to build the nuclear plant was a corporate mistake."[15] Despite Selby's return, the management changes continued. In November, A. H. Aymond—who had been chairman of the Consumers Power board from 1960 through 1978 and had remained on the board after he was replaced as president of Consumers Power in 1975 and also after he was replaced as chairman in 1979—resigned to give the utility's new management a "fresh start." "I believe with new management taking over at Consumers Power, it is an appropriate time to end my long association with the Company," he said.[16]

In the last days of Selby's, Boris's, and Aymond's tenures at Consumers Power, the company's financial condition was terrible. It was so close to defaulting on its major debt that it had to cajole General Motors to pay its September 1984 electric bill early so the utility could make a debt service payment.[17] Loss of bank lines of credit made it necessary for Consumers Power to borrow $240 million from its two major gas suppliers to finance seasonal natural gas inventories for the winter of 1984–1985.[18] In the fall of 1984, a long-term coal supplier broke with past practice and demanded nine advance payments of $190,000 each.[19]

When John Selby resigned in 1985, the utility desperately needed someone who could rally political support for whatever it would propose to do to recover from its devastating financial situation. As noted earlier by Mike Hayes's comment about Consumers Power's lack of effort to gain support for the Midland Plant in Lansing in 1980, Selby had not done anything to gain political support for Consumers Power. An illustration of Selby's lack of presence in the political arena outside of Jackson is that in the wake of his 1982 election, Governor James Blanchard had created a blue-ribbon panel consisting of twenty-seven business, civic, and labor leaders called the Michigan Financial Crisis Council to look at the state's budget deficit. The council was chaired by Walter J. McCarthy Jr., chairman of Detroit Edison. Selby was not even a member; at the time, he was a Jackson Community College trustee.[20] "[Selby] did things because he thought they were the right things to do, not because they were politically savvy," said John W. Clark, a Consumers Power senior vice president. "When the going got tough, we [Consumers Power] reached out for the political support, and it just wasn't there. The company had never gotten out in the state to demonstrate need for the plant."[21]

On September 25, 1985, the utility's board of directors selected Willian T. (Bill) McCormick Jr. to replace Selby as the new chairman of the board and chief executive officer. Unlike Selby, McCormick was extremely adept in the political arena, having started his career in Washington, D.C., after earning a doctorate in nuclear engineering from Massachusetts Institute of Technology. Between 1973 and 1976, he held several senior posts in the federal government, including policy level positions in the Energy Policy Office at the White House, the Office of Management and Budget, and the U.S. Energy Research and Development Administration. After Gerald Ford lost the 1976 presidential election, McCormick then worked for the American Gas Association as a lobbyist, first with the title of vice president for policy analysis and later as vice president of government relations.[22]

Once at Consumers Power, McCormick worked on recovering money from Bechtel and on solving the problem of what to do with the Midland Plant. The utility—which in 1987 became the primary subsidiary of a newly formed holding company McCormick had formed called CMS Energy Corporation—did recoup some money from the engineering firm. In 1987 Consumers Power sold a two-thirds interest in its Palisades Plant to Bechtel Power Corporation for $50 million in cash plus a payment of $400 million in cash in 1988. (The utility had spent $188 million in constructing the Palisades Plant.) Company spokesperson Paul Knopick commented, "This turns a bitter, major, lengthy, contentious lawsuit against Bechtel into a working arrangement that is beneficial to both companies."[23]

At the time of the sale, the amount in dispute over the Midland Plant was not revealed. A joint announcement from McCormick and Bechtel president Cordell Hull said only that the transaction was "an innovative resolution to our differences over Midland, and is one that provides both organizations with the opportunity for near- and long-term benefits."[24] It is not clear why Bechtel thought the purchase of a two-thirds interest in the Palisades Plant was a good deal, unless it was because Consumers Power agreed to buy the entire output of the plant under a long-term contract.[25] Bechtel must have believed that the problems that the Palisades Plant had could be managed, and with a long-term sales contract with Consumers Power for electricity produced by the plant, the engineering firm would realize a reasonable rate of return on its investment.

The cash Bechtel gave Consumers Power was what the utility needed to try to bring itself back from the horrible shape it was in after cancelling the Midland Plant.[26] Selling an interest in the Palisades Plant also helped Consumers Power get rid of a plant with a terrible history. The Palisades Plant, plagued by constant shutdowns and operating at less than half its capacity since its opening in 1971, had one of the worst performance records of any commercial nuclear plant in the country.[27] Although the settlement provided some relief, McCormick's major challenge was to figure out how to derive some value from the Midland Plant so Consumers Power could carry on.

▪ ▪ ▪

Meanwhile, Dow Chemical Company also had a problem in Midland. When the chemical company cancelled its contract with Consumers Power in 1983, it sued the utility for an unspecified amount in excess of $60 million. Consumers Power in turn sued Dow Chemical for a termination payment of $460 million.[28] And in 1984 Dow Chemical still did not have an adequate replacement for its coal plants. The situation in Midland was dire. Cal Ieuter, president of the Midland Chamber of Commerce in 1984, said, "ABATE and General Motors presented too big an obstacle . . . Outstate Michigan has lost. Gov. Blanchard let us down. The loss of the tax base will be the biggest loss . . . But Midland will survive and go forward."[29] Although in 1984, this declaration may have sounded like just a piece of Chamber of Commerce community booster talk, it actually proved to be true.

CHAPTER 23

Consumers Power to Consumers Energy

From Nuclear to Natural Gas and Beyond

In 1985 one writer said he thought the never-ending saga of the Midland Plant deserved a slot on daytime television. The story had everything a drama needs: high-stakes conflict over money, technology, politics, power, and—according to the Consumers Power Company—the future of Michigan's economic development hung in the balance.[1] Like a good soap opera, there's always another episode. This next episode begins in July 1985, when the Michigan Public Service Commission (MPSC) allowed Consumers Power a $94 million rate increase that it desperately needed. But in order to get it, Consumers Power had to promise not to spend any money directly or indirectly to restart construction on the Midland Plant without the MPSC's approval.

One of the first things Bill McCormick did was to order a study of what to do with the failed nuclear facility. Even though this was a logical step, the announcement of the study caused the MPSC to send Consumers Power a letter in October 1985 saying that it violated the terms of the agreement for the $94 million rate increase and if the utility proceeded it would have to forfeit the rate increase.[2] In November 1985 McCormick obtained grudging approval for the study from MPSC's chief of staff, Roger Fischer, as long as it was done by the utility's own staff.[3] Comments on the effort varied from skepticism to "hell no." Fischer said Consumers Power probably would have difficulty gaining MPSC's support for restarting the project. Hugh Anderson, assistant state attorney general, said, "[McCormick] is off to a bad start if he wants to complete the plant. He's fighting the last war."[4]

Nevertheless, the study went ahead. James Cook led the five-month investigation, called the Midland Options Study, from late 1985 to the spring of 1986. He had joined Consumers Power in March 1977 as vice president, energy planning.[5] In October 1980 he was given the title vice president projects, engineering and construction, with the major project under his direction the Midland Nuclear Plant.[6] According to the 2017 Consumers Power coffee table book, *Building for the Future: The Story of Consumers Energy 1886–2016*, the Midland Options Study used "sophisticated computer models and analytics."[7] It considered four options: (1) abandoning the facility, (2) converting to a coal-fired facility using some of the equipment from the nuclear design, (3) putting everything on hold and at a later date finishing the facility as a nuclear plant, and (4) converting to a natural gas-fired facility.

Consumers Power was not unique in its problem with what to do with a nuclear plant that could not be finished.[8] The first three possibilities in the Midland Options Study were eventually implemented by other companies. The Public Service Company of Indiana implemented option one. It decided on January 16, 1984, to stop building the Marble Hill nuclear plant because after spending $2.8 billion it could no longer raise money for construction. All equipment was sold, and the plant was demolished.[9] Cincinnati Gas and Electric accomplished option two, converting an abandoned nuclear plant to a coal-fired plant. At the time it ceased construction on the Zimmer Nuclear plant, the Ohio utility had invested about $1.6 billion.[10] The Tennessee Valley Authority (TVA) chose option three—putting everything on hold and finishing it at a later date—for its Watts Bar Unit 2 Plant. The TVA had started construction on this facility in 1973 with an estimated completion cost of about $400 million and had stopped in 1985 when it was approximately 60 percent complete at a cost of $1.7 billion. Construction did not resume until 2007, to be completed in 2015, with commercial operation commencing in 2016. Although TVA eventually succeeded in completing Watts Bar Unit 2, it illustrated that even in the 2000s it was not possible to complete nuclear plants free of difficulties with forecasting, construction management, regulatory requirement changes, and cost overruns. Because of these problems, in 2011 a new management team was established, and it took a number of steps that allowed the the plant to be completed.[11]

But the fourth option in the Midland Options Study was—and to this day is still—unique: convert the Midland Plant into one using natural gas for fuel. This plan made use of the steam turbines, diesel generators, cooling pond, electric switchyard, office buildings, and warehouse facilities that had been constructed for the nuclear version of the plant.[12] Thanks to the arrangement subsequently made with Dow Chemical, Consumers Power was also able to use the steam line to Dow Chemical that it had constructed for the nuclear plant. The utility's board of directors approved the gas conversion plan on April 8, 1986.[13] The design used natural gas turbines to produce electricity and to use the high-temperature gas turbine exhaust (which would otherwise all escape up the exhaust pipe) to make steam in a heat recovery unit. The company would then pipe it to a steam turbine left over from the nuclear plant, and the turning of the steam turbine would generate electricity.[14] The term for a facility of this type is a natural gas combined-cycle, cogeneration plant.[15] With the Midland Plant, Consumers Power had encountered crippling problems with funding, construction schedules, and building a plant to specifications. This conversion offered a number of construction advantages: less additional capital investment than conversion to coal, highest confidence in completion cost, shorter schedule for completion, and modular design that made for ease of construction and operation.[16]

Since McCormick had a business background in natural gas, it may not be surprising that option four was selected. After leaving his Washington, D.C., job with the American Gas Association in 1978, he had moved to Detroit and worked for American Natural Resources Company (ANR), a natural gas pipeline and storage company. His titles included chairman and CEO of ANR and executive vice president and director of its parent company, the Coastal Corporation.[17] However, McCormick's response to a question from Midland shareholder Susan Campbell at Consumers Power's annual meeting of April 8, 1986, would indicate that he was not necessarily married to the natural gas solution. Campbell asked, "Disregarding the lack of capital and the so-called anti-nuclear political arena . . . do you think Midland should have been completed as a nuclear facility?" McCormick's reply did not reveal any prejudice towards natural gas: "Political and regulatory uncertainties are obstacles to pursuing the nuclear option, but Consumers would have chosen the nuclear option if not for political opposition and prospective regulatory problems."[18]

McCormick needed the approval of the MPSC to proceed beyond the work of Consumers Power's staff with the plan for the natural gas conversion, and in 1986 he applied all of his political skills to convincing everyone he could that the gas plan was the answer. By the middle of May, he was able to announce support from

173 companies, sixty business organizations, nine labor groups, twenty-two local governments, the Republican and Democratic leaders of the state House and Senate, nineteen of Michigan's twenty members of Congress, and seventeen "miscellaneous" hospitals, service clubs, churches, and colleges.[19] The political blitz succeeded. In October 1986, the MPSC agreed that Consumers Power could spend up to $50 million of its own or borrowed money to develop the engineering details for the conversion of the Midland Plant according to the plan the utility had developed in the study.[20]

▪ ▪ ▪

In addition to gaining support for the conversion of the Midland Plant, McCormick had one other problem in 1986: what to do about the dueling lawsuits that had been filed by Dow Chemical and Consumers Power. On April 7 McCormick sent a letter out to the Michigan business community touting the conversion. In the copy of this letter he sent to Paul Oreffice, president and CEO of Dow Chemical, McCormick scrawled "P.S. Do you need any steam?"[21] It is not clear what exactly he intended by his handwritten note, but Oreffice took it as a serious question and sent the letter to Hunter Henry, head of Dow USA. Henry in turn sent the letter on to F. R. (Pete) Lehman, head of Dow Chemical's Michigan Division.[22] Lehman believed it was clear to him when he was given the letter from McCormick that he needed to "solve the problem."[23] It was time to make a deal.

Meetings between the parties started in mid-May 1986. McCormick told Lehman that Consumers Power was determined to convince the MPSC to approve the conversion of the Midland Plant. Discussions continued through the summer of 1986. By that fall Consumers Power and Dow Chemical had come up with the idea of forming a partnership to finance the $560 million that would be needed.[24] This agreement, signed on January 27, 1987, included a plan to install twelve large gas turbines on the site of the failed nuclear facility, which would be part of the equipment owned by the new partnership entity, Midland Cogeneration Venture Limited Partnership (MCV). On the same day, Midland County Circuit Court Judge David Scott DeWitt—after being told the companies had reached an agreement—dismissed the lawsuits to end the three-year-old litigation between Dow Chemical and Consumers Power.[25]

The original partnership MCV ownership structure provided for CMS Midland, Inc. (a subsidiary of CMS Energy) to be a 49 percent partner; Rofan Energy, Inc. (a subsidiary of Dow Chemical) to be a 10 percent partner; and for additional partners to be added.[26] The Consumers Power name was not part of the agreement because the company had been reincorporated as part of CMS Energy Corporation, a holding company with utility (Consumers Power) and nonutility (Northern Michigan Exploration Company) assets. The MCV plant design that CMS and Dow Chemical worked out called for steam to be extracted from the steam turbine used by the gas-fired plant and sent to the chemical company through the steam line that had been constructed for the nuclear design. The plan also called for electric transmission lines to run from MCV to Dow Chemical so that the gas-fired plant could supply electricity directly to Dow Chemical.[27]

Several other entities were added to complete the ownership in the MCV, including subsidiaries of gas pipeline companies with which MCV would sign gas transportation agreements (Panhandle Eastern Corporation and the Coastal Corporation) as well as subsidiaries of companies from whom MCV would purchase equipment and services to build the gas-fired plant (Asea Brown Boveri, Inc.; Combustion Engineering; and Fluor Corporation). By 1988 it seemed that the rules of the game were that if it wanted to provide services or equipment to MCV, a company would also acquire an ownership interest in the the venture.

Rodney Boulanger, neither a Consumers Power nor a Dow Chemical person, was selected as the first president and chief executive officer of the MCV in 1987. Prior to this time, he had been president of the ANG Coal Gasification Company that managed the Great Plains coal gasification plant in North Dakota on behalf of the U.S. Department of Energy.[28] Before that, he had worked in Detroit with McCormick at ANR.[29] Although McCormick first thought that the management of MCV would be located in Jackson, Michigan, where Consumers Power had its headquarters, Boulanger was successful in changing that plan so that MCV management was located in Midland. In 2018 he said that he had thought, due to the contentious history between Consumers Power and Dow Chemical, it was extremely important that MCV establish an identity as a completely independent organization with its own policies and procedures. McCormick also thought that Boulanger should function with a minimal staff with most of the work farmed out to employees of Consumers Power. On this issue Boulanger also prevailed and hired his own MCV staff. For his management team, in 1988 he hired no Consumers Power or Dow Chemical employees for legal, financial, human resources, electric marketing, or gas supply positions.

Consumers Power entered into an electric supply contract with MCV to purchase up to 806 MW of electricity in 1990, increasing in 1995 to 1,240 MW.[30] Under the January 27, 1987, Steam and Electric Agreement between Dow Chemical and MCV,

FIGURE 16. Dedication of the Midland Cogeneration Venture Limited Partnership's (MCV) gas-fired power plant, April 9, 1990. William T. (Bill) McCormick Jr., chairman of the board and chief executive officer of CMS Energy (new parent company of Consumers Power); James Blanchard, governor of Michigan; Bill Schuette, member of the U.S. House of Representatives from Michigan's 10th congressional district; and unidentified man tour the MCV plant. (© 1990 MLive Media Group. All rights reserved. Used with permission.)

the former was given the ability to take up to a maximum average hourly flow rate of 691,900 pounds of steam and an average hourly power rate of 60 MW of electric power for use at its Midland site (provisions existed in the contract to raise both the steam and electric quantities).[31]

Built on time and under budget, MCV first generated electricity commercially on March 16, 1990, utilizing eight gas turbines and one steam turbine; full operation of the facility with all twelve gas turbines occurred in the second quarter of 1990.[32] A celebration for the MCV's gas-fired plant was attended by over four hundred people on April 9, 1990, and no Michigan politician could miss it.[33] Present were Governor James Blanchard, State Senate Majority Leader John Engler, U.S. Senator Carl Levin, and U.S. Representative Bill Schuette (who was going to oppose Levin in the coming race for the U.S. Senate). Blanchard called the MCV an example of the state's development successes. Engler said the MCV represented an investment in courage and foresight. Levin said the plant represented the development of Michigan and protected the environment at the same time. Schuette said it was a powerful statement about Michigan's future. As for McCormick, he said, "Three years ago, conversion of an idled nuclear plant to natural gas was a dream—something that had never been done. To now have this power plant in operation ahead of schedule and under budget is an historic event."[34]

The successful conversion of the plant put the utility and the Midland area back on a sound financial footing. The MCV agreement had provided the company with interest bearing notes totaling $1.27 billion (see appendix 5 for values placed on equipment constructed for the Midland Plant), and in November 1989, CMS reestablished its common stock dividend, the first dividend since August 1984.[35] After 1990, the holding company, Consumers Power Company (later to be called Consumers Energy), and MCV would continue to haggle with Attorney General Frank Kelley, ABATE, and others in both court cases and cases before the MPSC over finance arrangements, power supply contracts, and gas supply arrangements.[36] Nevertheless, as of 2018 MCV had made a major economic impact, paying local and state taxes of $460 million for years 1990–2018, providing direct employment for approximately 250 full-time equivalent positions, selling 1,240 MW of capacity and energy to Consumers Energy Company, selling up to a power rate of 150 MW of electricity and 1,000 kp/hr of steam to the Dow Chemical, and selling up to 250 kp/hr of steam to Dow Corning Corporation.[37]

■ ■ ■

In 1971, fifteen thousand people attended a rally to show their support for bringing nuclear power to their own backyards. The words to the song they sang that sunny October day called for nuclear power to provide energy for their jobs, cleaner air, and cleaner water. Today the coal mess that was a blot on the Midland landscape in 1967 no longer exists because of the switch from Dow Chemical's coal plants to MCV's natural gas plant. But to many the problem with natural gas is that, although it does not cover the city in ash, it only reduces the greenhouse gas emissions of carbon dioxide to approximately half of those of coal.

In 2021 those who want electricity generated with zero carbon emissions appear to be winning the high stakes game for energy supply. In 2019 Consumers Energy received approval from the MPSC to implement its integrated resource plan for a major change in its electric generation by 2040.[38] The biggest change planned is a massive addition of solar and wind generation to increase the supply of renewable, zero-carbon sources from 11 percent in 2019 to 56 percent in 2040.[39]

Not since Consumers Power's 1970s vision that with the addition of the Midland and Quanicassee plants they would have 45 percent of their generating capacity from nuclear by 1980 has the utility proposed such a radical change.[40]

Consumers Energy is also planning other major changes in its electric generation. In 2019, 31 percent of the supply was from natural gas, and the goal is to reduce that to 10 percent by 2040. In 2019, 28 percent came from coal and nuclear sources, and the goal by 2040 is to eliminate these. The Consumers Energy that Patricia Kessler Poppe, president and CEO of its parent company CMS Energy, envisioned in 2019 for 2040 was certainly not her father's Consumers Power.[41]

Not everybody is on board with Consumers Power's plan for 2040. Michael Moore, who marched in opposition to the Midland Plant in 1979, is the executive producer of the documentary *Planet of the Humans* released in 2019.[42] The general thrust of his new film is a pox on all means of electricity generation because we humans are destroying our planet with our addiction to energy. The conclusion drawn is that even the construction and use of renewable generation destroys our planet. Should his viewpoint be adopted, people will no doubt soon be demonstrating against the construction of new wind and solar facilities.

Indeed, Consumers Energy's plan for 2040 will require acres and acres of new solar arrays and wind farms. Matching just the 1990 MCV gas-fired generating capacity of 1,300 MW in 2020 with solar will require approximately eight thousand acres of Michigan countryside be transformed into an area with the aesthetic appeal of an unfenced auto junkyard. Building massive new wind farms will bring the wappty-wappty noise of windmills to the backyards of many rural Michigan residents.[43] This is all a far cry from Consumers Power's 1969 claim that their new nuclear generation would not be placed in unspoiled rural areas. Furthermore, new transmission and distribution lines will need to be built to transport electricity from new generation facilities to customers. All of this new construction will not go without opposition; a lot of people just don't want anything new in their backyards. And the 2040 plan assumes that 22 percent of the electricity needed in 2040 will come from electric customers' "efficiency programs." Will customers really be willing to give up control of their heating and air conditioning to Big Brother?

Some will point out that in cloudy and snowy Michigan, solar generation does not make good economic sense since the cost of electric supply and storage backup for intermittent power from solar farms when the sun is not shinning or they are covered with snow has not been properly calculated. Although the Great Lakes provide the best locations for wind generation, many with their multimillion dollar homes on the shoreline will oppose locating wind turbines in the lakes because they won't want to look at them.

With the reduction of natural gas electric generation—which is available around the clock—in favor of intermittently available wind and solar, how much would Consumers Power be forced to pay if it has to buy electricity in the open market and purchase expensive storage equipment to provide electricity when the sun is not shining and the wind is not blowing? When a wave of polar air gripped the Midwest on January 31, 2019, the expectation that wind generation would be available to meet record electricity demand proved wrong. Not only did the wind quit blowing as a result of the cold wave, it turned out the wind turbines needed to be heated because of the sensitivity of their electric cabinets, gearboxes, generators, and lubricants. In one case, wind farms went from being a power producer to a 2 MW user in order to heat their equipment.[44] This situation resulted in a 6 gigawatts generation deficit throughout the morning. Consequently, energy costs throughout the Midwest rocketed from $34/MWh at 6 a.m. (CST) to $578/MWh by 9 a.m. (CST).[45]

▪ ▪ ▪

Given the cost of backup electricity for all of the new intermittent facilities that are planned, will it be possible to deliver electricity at a price that is attractive to industry? Like Consumers Power's plan in the 1970s to add large amounts of new nuclear generation, its 2019 integrated resource plan is a financial bonanza for solar and wind contractors and an opportunity for the utility to add to its rate base. But in the Wonderland of utility regulations, the utility customer does not pay based on the ability of a utility to deliver the lowest cost electricity, but on what the MPSC approves.

The opposition to Consumers Power's plan for 2040 will probably not be as focused as it was against the Midland Plant, but it is likely that a complex array of opposition forces will mobilize as the utility seeks to execute its latest plan to provide electricity for its Michigan customers. The results of the high-stakes conflict among money, technology, politics, power, and the future of Michigan's economic development will, as it did for the Midland Plant, determine the price of power for the future.

Epilogue

What can be said about Midland in 2020? At the beginning of this story I talked about the uneventful 1967 start-up of a small nuclear reactor in Midland. In 2020 Dow Inc. continued to operate this small research reactor. Low-power reactors, like Dow's, are used for research and development purposes. This valuable research tool assists the company's global businesses by reducing the time associated with product development and performance evaluation. The reactor undergoes all required inspections and is in full compliance with regulatory authorities.[1]

The new Dow Inc. no longer employs twelve thousand people in Midland as Dow Chemical did in 1967. A job count at the end of 2016 was eight thousand.[2] The estimated population in the City of Midland in 2017 was forty-two thousand, and therefore Dow Chemical remained a major employer.[3] However, direct comparisons between 1967 and after 2017 are not possible because of the major corporate changes that occurred in the latter year. On September 1, 2017, Dow Chemical and E. I. du Pont de Nemours & Company (DuPont) announced their successful merger into a company called DowDuPont. On February 26, 2018, this new corporate entity announced the creation of yet another new company called Dow Inc., which is based in Midland. The focus of this firm is sustainable materials science. The new

Dow Inc. separated from DowDuPont on April 1, 2019, to once again become an independent, publicly traded company. As of 2020, Dow Inc. said it employed around five thousand employees and four thousand contractors in Midland. Its stated ambition is to be the most innovative, customer-centric, inclusive, and sustainable materials science company in the world.[4]

▪ ▪ ▪

Midlanders are still proud of their city and extol its virtues. A local organization, Mid-Century Modern Midland, celebrates Midland's architectural heritage. Its website details an amazing four hundred mid-century modern structures woven throughout the community.[5] The Herbert H. and Grace A. Dow Foundation and other foundations in town continue to provide funds to make it a nice place to live. In 2018 the nation's longest canopy walk opened in the Whiting Forest in the middle of town, part of a $20 million project to redevelop the wooded area.[6]

Although the Black population is still only about 2 percent, on June 7, 2020, over one thousand people marched on Saginaw Road in town in favor of social justice in a Black Lives Matter demonstration.[7] This predominantly White group marched in spite of the fact that their community had just experienced a devastating flood when dams failed on the Tittabawassee River, and most wore masks to protect themselves and others from the COVID-19 pandemic.[8]

Midland still reveres chemistry and innovation. The Ned S. Arbury pocket park at one end of the downtown Main Street business section displays a bronze casting celebrating Midland as the "City of Modern Explorers," and a decorative iron screen that depicts the repeating pattern of a polystyrene molecule (whose design was suggested to the artist by Dr. Etcyl Blair, a retired Dow Chemical chemist).[9] The other end of the downtown business district on Main Street features a small park with monuments commemorating the chemical reactions of early Dow Chemical products. Midland High School's symbol is still the atom, and its students still call themselves the Chemics. The highway entrances to Midland are now marked with large copper medallions inscribed "City of Modern Explorers."

In 2021 it is not clear whether innovation in Midland will continue to be assisted by the Midland Cogeneration Venture Limited Partnership (MCV). The future of MCV to provide jobs and tax revenue for Midland will depend on its ability to continue to contract for the sale of steam and electricity it produces from natural gas. This ability will depend on whether low-cost natural gas continues to be made possible by hydraulic fracturing techniques that are used to recover it. The purchase of steam and electricity produced by burning natural gas will depend on whether utilities and industrial users of steam and electricity decide that natural gas is an acceptable fuel. In 2021, it appears that Consumers Energy will continue to purchase electricity from MCV at least through 2030 based on an order published by the Michigan Public Service Commission on March 4, 2021.

Radioactive Waste Concerns

When I interviewed Charles Overberger in 2017 and 2020, he did not remember exactly what the source of his information about the pollution was that he had mentioned in his 1971 interview. However, he said his mother was a friend of Mary Sinclair so his information could have come indirectly from her. That year Sinclair circulated a revised copy of her pamphlet, *Nuclear Power & Public Concern*, in Midland. It consisted of short quotations from a large number of sources about concerns with various aspects of nuclear power. A sample of the passages on radioactive wastes and radiation follows:

> at the Hanford atomic products operation 140 huge tanks, each 50 feet high, are buried near the center of the reservation. These tanks contain some 50 million gallons of highly radioactive waste liquid, some of it boiling hot, the waste product of nine nuclear reactors . . .
>
> Even without accidents or earthquakes, tank storage isn't necessarily safe. Radioactive wastes corrode tanks; by the year 1965, four had developed leaks. Should another one possibly go undetected, the nearby Columbia River could become deadly.[1]
>
> . . .

> When I was chairman of the AEC from 1946 to 1950, highly poisonous radioactive wastes were being produced at the AEC reactor. During the years since the AEC, has the problem of disposition of these poisonous wastes been made demonstrably manageable, either as the danger or as to cost? The answer is No![2]

> Permissible limits as set by the Atomic Energy Commission should not necessarily be construed as safe levels. The difficulty of setting these levels is that there is no adequate sample size to derive data from. Numbers are all supposition as yet because no one yet knows the rate of cancer produced by a given dose of radiation but we have reason to expect that there is no safe dose of additional exposure.[3]

Although the details of waste storage at the Midland Plant site were probably not known in 1971, an article published in the *Detroit Free Press* on April 21, 1979, said:

> Each reactor at the Midland facility will have about 90 tons of uranium in its core. About one-third of this fuel will be replaced annually, leaving the company with the problem of storing 60 tons of spent, but still highly radioactive, fuel each year. The solution at Midland has been to build 50-foot-deep concrete-and-steel-lined pits inside an auxiliary building. The spent fuel bundles will be stored in these pits under boron-impregnated water. Wally Sabo, a Consumers construction engineer, said the storage pools would suffice through 1990, "and we hope the fuel processing plant will be running again by then."[4]

In 1979 some people assumed that fuel processing plants would become operational in the future and that radioactive waste from the Midland Plant and other operating nuclear facilities could be moved there from their storage pools. However, these processing facilities did not start operating, and the other option for the underground storage of nuclear waste at a central repository has not been completed as of 2021. At the time of this writing, nuclear waste was stored at the site of the nuclear plant. This means that, had the Midland Plant been completed, in 2021 its nuclear waste would be stored there on the banks of the Tittabawassee River. With no answer as to the cost for the permanent storage of nuclear waste, it is impossible to know the cost of nuclear power.

Nuclear Reactor Types

The world's first human-made nuclear reactor, Chicago Pile-1, used graphite to slow down or moderate the speed of neutrons being emitted from the element uranium to facilitate the collision of these neutrons with other atoms of uranium and keep the fissioning going. Slowing down some of the neutrons facilitated the collision of these first neutrons with other atoms of uranium. With the addition of the fissioning caused by allowing more collisions of neutrons with atoms of uranium it was possible for Fermi to create a chain reaction of fissioning. The amount of free energy available from the fissioning of the core of atoms is millions of times the amount of free energy available in a similar mass of chemical fuels like coal or natural gas, which rely on normal chemical reactions.

In Fermi's reactor no radiation shielding or cooling system was necessary since it operated at extremely low power. Chicago Pile-1 had control rods that could be inserted into the reactor to absorb neutrons and stop the chain reaction. In an atomic bomb, by contrast, there is no mechanism to stop nuclear fissioning once it starts. The entire force of the nuclear reaction is released almost instantaneously in the form of heat and energy. In addition radioactive products, often called ionizing radiation, are released. This ionizing radiation is extremely hazardous for all living plant and animal cells.

In contrast to the 1942 Pile–1 reactor, the breeder reactor used in the 1952 Fermi 1 plant used liquid metal (sodium) to transfer heat away from the reactor. In light water reactors, ordinary water is used as a moderator and to transport heat from the reactor. The term "light water" differentiated this reactor design from one called "heavy water," which used a rare compound composed of oxygen and a form of hydrogen called deuterium to act as a moderator and to transfer heat from the core. Unlike ordinary water with its familiar chemical composition of H_2O, heavy water includes two atoms of deuterium instead of two atoms of hydrogen. In addition to the differences in their moderating and cooling mechanisms, Pile-1, breeder reactors, and light water reactors all use different blends of nuclear fuel.

Westinghouse and General Electric took the lead in designing and manufacturing two types of light water nuclear reactors in the United States. Westinghouse's maintained the water in the reactor core under high pressure; this was called a light-water pressurized reactor (PWR). General Electric's did not maintain the water at high pressure and was called a light-water boiling water reactor (BWR).

Later Babcock & Wilcox and Combustion Engineering joined the competition. Hot water from a PWR manufactured by Babcock & Wilcox transferred heat from the reactor into a heat exchanger. Heat was transferred from the heat exchanger in the form of steam to generate electricity and then went through another heat exchanger to provide steam to Dow Chemical Company. (Appendix 3 shows a simplified diagram of a Midland Plant Reactor.)

Simplified Diagram of the Midland Plant

The Midland Plant was designed so the nuclear reactor located inside a containment building would generate heat, which would heat water under pressure to be pumped to a steam generator. The heat coming from the reactor would be transferred in the steam generator to pipes where steam would be generated and then sent to the steam turbine. The steam turbine would turn an electric generator to make electricity. The water in the pipe loop coming from the reactor would never touch the loop of pipe taking steam to the steam turbine. In order to produce steam for Dow Chemical Company, steam was to be extracted from the steam turbine and sent to a process steam generator.

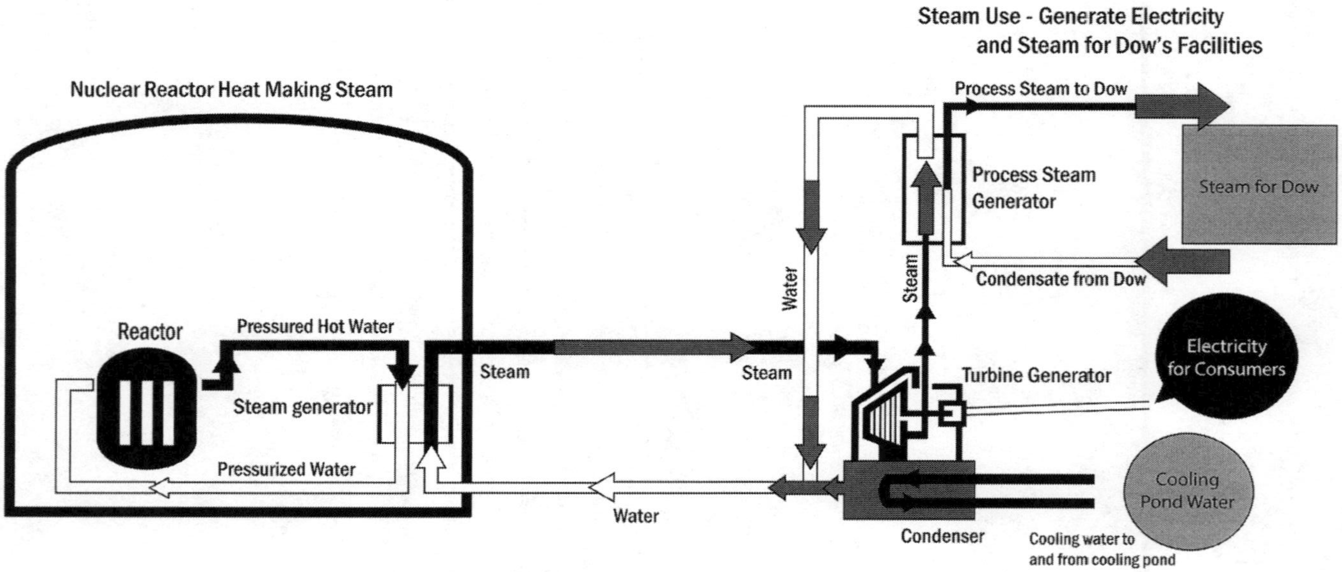

Simplified Diagram of the Nuclear Midland Plant

FIGURE 17. This simplified summary diagram of the operation of the pressurized water reactors installed at the Midland Plant and the general layout of the Midland Plant is based on the author's observations of the Midland Plant site and the following sources: David Bodansky, *Nuclear Energy*, 2nd ed. (New York: Springer-Verlag, 1996); Allan Sloan and Kirk Cheyfitz, "Nuclear Power Dream Turns Sour, *Detroit Free Press*, April 13, 1975, 1A, 12A; and Guy Snyder, "Nuclear Power Plant Const. 30% Complete at Midland," *Michigan Contractor & Builder* (Detroit), September 10, 1977, 8.

Fix for Cracking in the Walls of the Auxiliary Building

The auxiliary building, situated between the two reactor containment buildings, housed safety-related equipment including the control room, the access control room, cable spreading rooms, engineered safeguard systems, switchgear equipment, steam lines, and feed water pipes, as well as the facilities for handling, storage, and shipment of nuclear fuel. Although most of the auxiliary building sat on a concrete mat foundation poured below the level of the floodplain, the control room, two electrical penetration areas that housed cabling systems, and the feed water isolation valve pit were all originally supported by fill.

Initially Consumers Power Company argued that it was not necessary to take corrective action on the auxiliary building. The utility's experts testified in an Nuclear Regulatory Commission (NRC) hearing on December 1, 1981, that cracking in the building's cement walls was due to normal changes in concrete and that the cracks did not indicate that the walls were not in a sound condition. Executives said in January 1982 that the cracking was not due to the improper placement of fill dirt. "There has been no abnormal or unexpected settlement of any portion of the auxiliary building," stated James Cook, the company's vice president for projects, engineering, and construction. But in 1982 the utility eventually reached

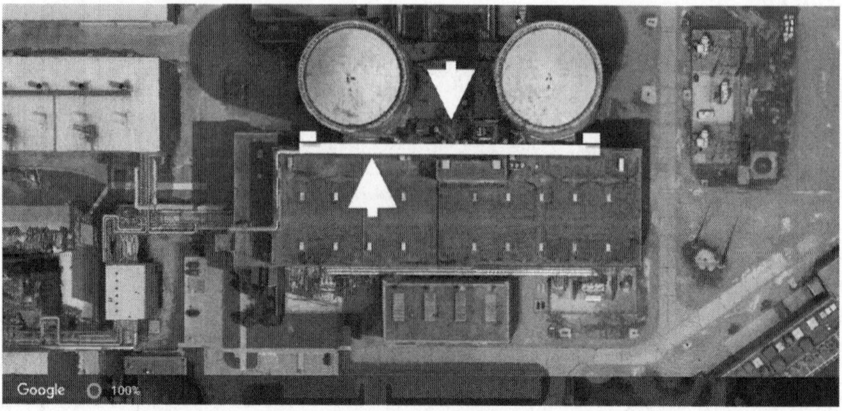

FIGURE 18. In this overhead view, the steam turbine building is the large rectangular building in the center of the picture. The nuclear containment structures appear as two large circles. Fill dirt was removed from under the auxiliary building (*down arrow*) situated between the two containment structures. A tunnel (*up arrow*) was dug underneath the steam turbine building to get to the area where the fill dirt needed to be removed. The tunnel was 275 feet long and was accessed from two shafts sixty-three feet deep (two white rectangles at ends of tunnel). (Courtesy of Google Earth/Landsat/Copernicus with notations added by author)

an agreement with the NRC to do the elaborate underpinning to place the auxiliary building on clay glacial till.

The tunneling and excavation were an astounding piece of work, and the preparation for them was also elaborate. Because of the relatively shallow water table in the excavation area, it was necessary to keep water from flowing in by constructing a freeze wall. This was accomplished by drilling a series of holes forty-four feet deep every four feet around the area and installing a closed loop system to circulate refrigerant to freeze the soil. Diagrams published showing the location of the freeze wall indicate that it was over one thousand feet in length.

Once the freeze wall was in place, workers did extensive tunneling, followed by removing fill dirt and installing piers. They started the tunneling by digging two sixteen by twenty-six foot shafts to a depth of sixty-three feet at each end of the turbine building to obtain access to the area where the work would be done under the six-foot thick floor of the turbine building. They connected the two shafts by digging a 275 foot-long tunnel. From the tunnel, they dug holes, and then fifty-seven concrete piers were poured resting directly on the clay glacial till to support the weight of the portion of the auxiliary building called the control tower, as well as the electrical penetration areas and the turbine building. They did much of the digging by shoveling dirt into buckets—two cubic feet at a time.

Once the piers were in place, they installed I-beams from the piers to the ledge of concrete mats under the containment buildings, and where the I-beams rested on the piers, they installed jacks. They raised these jacks to transfer the auxiliary building's weight to the piers. At this point, the workers accomplished a massive excavation called "the ballroom" by removing soil—enough to fill a football field fifteen feet deep. Inside the ballroom, they poured a permanent concrete wall up to twenty feet thick, thirty-five feet high, and 275 feet long. Then they lowered the jacks and transferred the weight of the auxiliary building from the fifty-seven temporary piers to the wall, which then became the permanent support that placed the weight of the structure directly on an undisturbed layer of clay instead of on fill dirt. On their way out, they backfilled the access shafts with earth.

In 1982 Consumers Power complained that actually the underpinning program had largely been necessitated because the NRC had ratcheted up their seismic requirements for nuclear plants. Cook said in a January 1982 article in the *Midland Reactor* that this massive effort was required primarily to meet the new higher NRC earthquake design criteria imposed in October 1980. The change meant that the magnitude of the underpinning program went far beyond Consumers Power's 1979 plan to correct potential soils deficiencies caused by improperly compacted backfill material.

Changing seismic criteria for nuclear plants is one example of how the NRC ratcheted up the regulations for nuclear plants during the time the Midland Plant was under construction. Criteria were drastically increased in 1976, and in 1978 seismic standards continued to be under review. Under the standards imposed in October 1980, the Midland Plant structures were required to be able to survive an earthquake of magnitude 5.0—far greater than any that had ever occurred in the Midland area and a standard 50 percent more stringent than before.

Conversion of the Midland Plant to a Natural Gas Plant

The gas-fired Midland Cogeneration Venture Limited Partnership Plant was designed so that electricity would be generated by twelve gas turbines, and the exhaust from its gas turbines passed over pipes in a heat recovery unit to generate steam. This steam was then piped to the steam turbine left over from the nuclear Midland Plant. As would have been the case in the nuclear plant, the steam turbine would generate electricity, and some of the steam in the steam turbine would be extracted to provide steam to Dow Chemical Company. In 1990 the total electric capacity of the plant was stated as 1,300 MW.

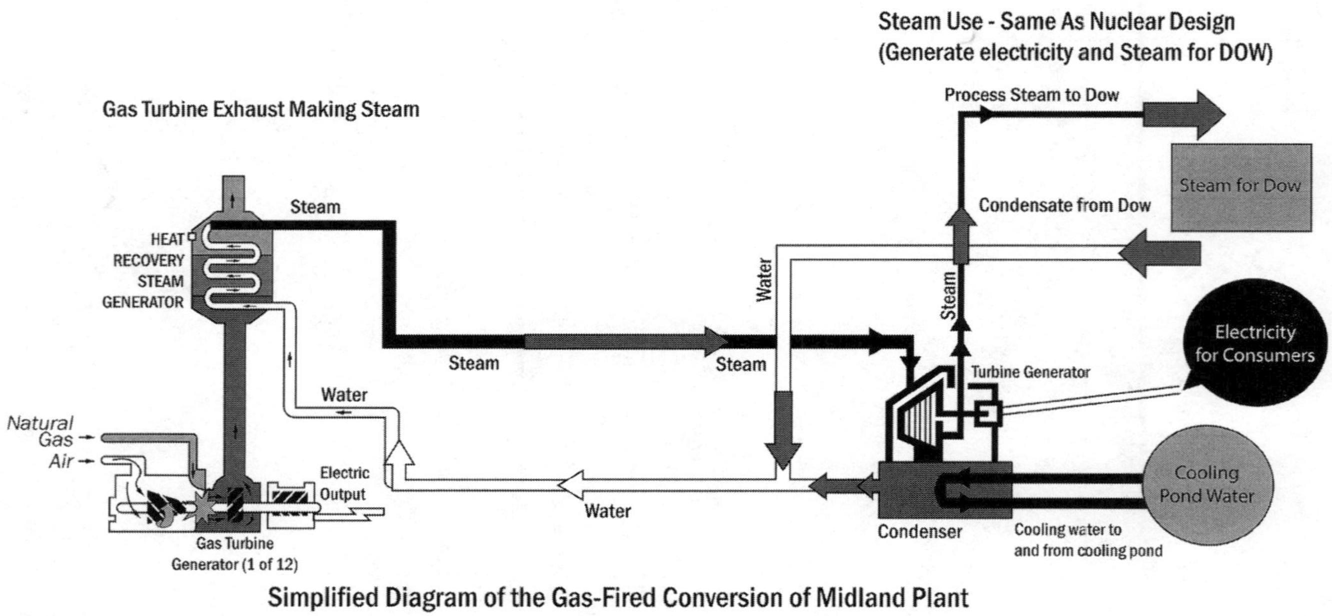

FIGURE 19. Simplified summary diagram of the operation of the gas turbines at the gas-fired Midland Cogeneration Venture Limited Partnership Plant (MCV) and the general layout in 1990 is based on the author's observations of the Midland Plant site and the following sources: C. E. Gatlin Jr., Gerald C. Vellender, and James A. Mooney, "Repowering of the Midland Nuclear Station," paper presented at the Tenth Annual Industrial Energy Technology Conference, Houston, TX, September 13–15, 1988, http://hdl.handle.net/1969.1/92393; Dow Chemical and Consumers Power, *Midland Cogeneration Venture* (Midland, Michigan: Dow Chemical and Consumers Power, April 1990), 3; and Guy Snyder, "Nuclear Power Plant Const. 30% Complete at Midland," *Michigan Contractor & Builder* (Detroit), September 10, 1977, 8.

FIGURE 20. The twelve stacks on the left are for the gas turbines installed by the MCV on the site of the Midland Plant. On the right, behind the steam turbine building, are two large concrete silos that were constructed for the two nuclear reactors of the Midland Plant. The MCV has now been in operation for thirty years. (Courtesy of Midland Cogeneration Venture Limited Partnership)

TABLE 5. Summary of sunk cost investment (in billions of dollars).

Reactor facilities and fuel	$1.6
Auxiliary systems	1.1
Diesel generator facilities	0.1
Turbine generator facilities	0.7
Process steam facility	0.2
Balance of plant	0.4
Total sunk cost	*$4.1*

Consumers Power Company, *Consumers Power Company: Beginning Our Second Century Powering Michigan's Progress* (Jackson, MI: Consumers Power Company, 1986), A1.24. This table was part of a notebook that Consumers Power circulated in Bill McCormick's 1986 campaign to gain support for the company's plan to convert the Midland Plant into a gas-fired cogeneration facility. According to George White, Consumers Power received $1.27 billion in interest-bearing notes and invested $1.5 billion in plant assets to receive a 49 percent interest in the Midland Cogeneration Venture Limited Partnership. George White, "Consumers Dow OK Midland Gas Plan," *Detroit Free Press*, January 28, 1987, 33 (and corrected *Detroit Free Press*, January 29, 1987, 3).

Notes

CHAPTER 1. LOCATION, LOCATION, LOCATION: GEOLOGY IS DESTINY

1. See, for example, Jacques-Nicolas Bellin, *Partie occidentale de la Nouvelle France ou du Canada* (Nuremburg, 1755), map. This map was published in 1755 by Homann Heirs. Library of Congress Geography and Map Division Washington, D.C., https://www.loc.gov/resource/g3310.ar001900/.
2. Paul D. Howell and Ben A. van der Pluijm, "Structural Sequences and Styles of Subsidence in the Michigan Basin," *GSA Bulletin* 111, no. 7 (July 1999): 974–991, 947.
3. George V. Cohee, "Geologic History of the Michigan Basin," *Journal of Washington Academy of Sciences* 55, no. 9 (December 1965): 211–233, 211.
4. E. N. Brandt, *Growth Company* (East Lansing: Michigan State University Press, 1997), 13, 33.
5. E. N. Brandt, *We Called It Mag-Nificent: Dow Chemical and Magnesium, 1916–1998* (East Lansing: Michigan State University Press, 2013), 4.
6. J. R. Stark and M. G. McDonald, *Ground Water of Coal Deposits, Bay County, Michigan, Geological Survey of Michigan: Michigan Technological University no. Open-File Report 80-591* (Lansing, MI: U.S. Geological Survey, 1980), 5.
7. Frank J. Krist Jr. and David P. Lusch, "Glacial History of Michigan, U.S.A.: A Regional Perspective," *Developments in Quaternary Sciences* 2, part B (2004): 111–117.

8. Brandt, *Growth Company*, 85.
9. Brandt, *We Called It Mag-Nificent*, 67.
10. "Text of H. D. Doan's Address at Stockholders' Meeting," *Midland Daily News*, May 3, 1967, 11, section 2.
11. "Dow Nuclear Reactor Start-up Successful," *Midland Daily News*, August 10, 1967, 1.
12. Jordon P. Howell, "Powering 'Progress': Regulation and the Development of Michigan's Electricity Landscape," *Annals of the Association American Geographers* 101, no. 4 (2011): 962–970, DOI: 10.1080/00045608.2011.569661. This article describes how the general notion that building of electricity infrastructure indicated socioeconomic progress was the prevailing view of both Michigan utilities and the Michigan regulator of the utilities, the Michigan Public Service Commission, dated back to 1915 and only started to be questioned in the 1980s.
13. "Civilian Unemployment Rate," Federal Reserve Bank of St. Louis, https://fred.stlouisfed.org/series/UNRATENSA#. "Average Hourly Earnings of Production and Nonsupervisory Employees, Total Private," U.S. Bureau of Labor Statistics, https://fred.stlouisfed.org/data/AHETPI.txt.
14. Consumers Power Company, *1967 Annual Report* (Jackson, MI: Consumers Power Company, February 16, 1968), 2.
15. Consumers Power Company, *1967 Annual Report*, 5, 6.
16. Consumers Power Company, *1967 Annual Report*, 12.

CHAPTER 2. IMAGINE THIS: A WHOLE TOWN RALLIES IN SUPPORT OF A NUCLEAR POWER PLANT

1. "Dow Midland Employees Live in 2,500 Square-Mile Area," *Midland Daily News*, July 7, 1967, 3A. Jerry M. Flint, "Midland Mich., Wants Nuclear Power," *New York Times*, December 3, 1971, 24.
2. "Consumers to Build $267 Million Nuclear Plant for Dow Power Use," *Midland Daily News*, December 14, 1967, 1.
3. Willis L. Shrink Jr., "Atoms for Peace in Pennsylvania," *Pennsylvania Heritage Magazine*, 35, no. 2, Spring 2009, 1.
4. Gerard H. Clarfield and William M. Wiecek, *Nuclear America: Military and Civilian Nuclear Power in the United States, 1940–1980* (New York: Harper & Row, Publishers, 1984), 277.
5. Irvin C. Bupp and Jean-Claude Derian, *Light Water: How the Nuclear Dream Dissolved* (New York: Basic Books, Inc., 1978), 49.

6. Bupp and Derian, *Light Water*, 49.
7. George Bush, *Future Builders: The Story of Michigan's Consumers Power Company* (New York: McGraw-Hill Book Company, 1973), 417–418.
8. "Witness of Tomorrow," editorial, *Brinewell*, January 1968, 1. *Brinewell* was a monthly publication of the Midland Division of the Dow Chemical Company, which is available from the collection of the Grace A. Dow Memorial Library in Midland, Michigan.
9. Julie Morris, "Thousands Rally in Midland To Back Building of A-Plant," *Detroit Free Press*, October 13, 1971, 3.
10. Morris, "Thousands Rally in Midland To Back Building of A-Plant," 3.
11. *Nuclear Power Rally Program* (Midland: Midland Chamber of Commerce), October 12, 1971, document from the collection of the Midland County Historical Society. Fat Bob had gained his fame on the highest-rated radio show in the Midwest, J. P. McCarthy's drive-time radio shows on WJR-Detroit.
12. "Fat Bob the Singing Plumber," *Recordo Obscura*, May 15, 2009, http://recordoobscura.blogspot.com/2009/04/fat-bob-singing-plumber.html.
13. Claudia Protz, "DC Meeting Arranged on N-Plant," *Midland Daily News*, October 13, 1971, 1.
14. Art Linkletter, *Kids Say the Darndest Things!* (Englewood Cliffs, NJ: Prentice-Hall, Inc., 1957). Although published in 1957, Linkletter's book was well known at the time he appeared in Midland in 1971. There continued to be enough interest in Linkletter and his book that a slightly different form was published in 2005 by Celestial Arts, an imprint of the Crown Publishing Company, a division of Random House. The 2005 edition was illustrated by Charles M. Schulz (a famous American cartoonist and creator of the comic strip *Peanuts*). A film clip of Linkletter speaking at the Speak Up Now rally was included in *Sixty Minutes*, "Mary and Goliath," produced by Christine Weicher, aired January 27, 1985, Columbia Broadcasting System. To view this recording, see "60 Minutes: Midland Nuclear Hearing, Part 1," YouTube video, https://www.youtube.com/watch?v=cbQso2PC6Jc; and "60 Minutes: Midland Nuclear Hearing, Part 2," YouTube video, https://www.youtube.com/watch?v=Tn_PiPHBXMg.
15. Protz, "DC Meeting Arranged on N-Plant," 1.
16. Midland Chamber of Commerce, *Nuclear Power Rally* (program for Nuclear Power Rally, October 12, 1972, Midland County Fair Grounds), Midland County Historical Society Files.
17. *Sixty Minutes*, "Mary and Goliath."
18. D. A. Phillips, "Speak Up Now Rally Was Double Attraction for 1,000-Plus Youths," *Midland Daily News*, October 13, 1971, 1.

19. Charlies Overberger, interview with author, 2017. Appendix 1 provides a summary of the concerns expressed in Midland about radioactive waste in 1971.

CHAPTER 3. REGULATORS: SAFETY RULES AND THE SPANISH INQUISITION

1. Samuel Walker and Thomas R. Wellock, *A Short History of Nuclear Regulations 1946–2009* (Washington, D.C.: U.S. Nuclear Regulatory Commission, October 2010), NUREG/BR-0175, Rev. 2, 1–2.
2. Constance E. Cook, *Nuclear Power and Legal Advocacy* (Lexington, MA: Heath and Company, 1980), 49.
3. Walker and Wellock, *A Short History of Nuclear Regulations 1946–2009*, 28.
4. Walker and Wellock, *A Short History of Nuclear Regulations 1946–2009*, 33–37. By the late 1960s a reactor core meltdown caused by a loss-of-coolant accident was seen as an unlikely but plausible accident. Experiments run in late 1970 and early 1971 suggested that the emergency core cooling systems (ECCS) for light-water reactors might not work as designed. In 1972 and 1973 the AEC held public hearings on the ECCS in light-water reactors, but these hearings only undermined public confidence in the safety of nuclear power.
5. Gerard H. Clarfield and William M. Wiecek, *Nuclear America: Military and Civilian Nuclear Power in the United States 1940–1980* (New York: Harper and Row Publishers, 1984), 55. Clarfield and Wiecek's estimates of deaths at Hiroshima and Nagasaki are from Eisei Ishikawa and David L. Swain, trans., *Hiroshima and Nagasaki: The Physical, Medical, and Social Effects of Atomic Bombings; By the Committee for the Compilation of Materials on Damage Caused by the Atomic Bombs in Hiroshima and Nagasaki* (New York: Basic Books, 1981), 364–369. Estimates of the total number of deaths vary. Seren Morris puts the estimated deaths at around 140,000 of Hiroshima and around 74,000 people in Nagasaki with the comment that due to the massive destruction of the cities recorded death tolls are only an estimate. Seren Morris, "How Many People Died in Hiroshima and Nagasaki," *Newsweek*, August 30, 2020. https://www.newsweek.com/how-many-people-died-hiroshima-nagasaki-japan-second-world-war-1522276.
6. Wilfred Burchett, "Chapter 1: The Atomic Plague [1945]," in *Rebel Journalism: The Writings of Wilfred Burchett*, ed. George Burchett and Nick Shimmin (Cambridge: Cambridge University Press, 2007), 1–5.
7. Richard Tanter, "Voice and Silence in the First Nuclear War: Wilfred Burchett and Hiroshima," *Asia-Pacific Journal* 3, no. 8 (August 3, 2005), http://apjjf.org/-Richard-Tanter/2066/article.html.

8. Boyce Rensberger, "The Stunted Children of Hiroshima," *Detroit Free Press* September 11, 1967, 13. James W. Wood, Kenneth G. Johnson, and Yoshiaki Omori, "In Utero Exposure to the Hiroshima Atomic Bomb," *Pediatrics* 39, no. 3 (March 1967): 385–392, https://pediatrics.aappublications.org/content/39/3/385.
9. Walker and Wellock, *A Short History of Nuclear Regulations 1946–2009*, 28.
10. "Mrs. Sinclair Tells Why N-Plant Hearing Recessed," *Midland Daily News*, August 21, 1971, 1.
11. Walker and Wellock, *A Short History of Nuclear Regulations 1946–2009*, 45.
12. Technical Information Service, *Information Report To the Project Companies of the Dow Chemical-Detroit Edison and Associated, Nuclear Power Development Project* (Oak Ridge, TN: Technical Information Services, 1953), 11.
13. "U.A.W. and the Auto Industry, May 1950," *New York Times*, October 8, 2015, https://www.nytimes.com/interactive/2015/10/08/business/uaw-auto-union-timeline.html#/#time386_11129.
14. John G. Fuller, *We Almost Lost Detroit* (New York: Reader's Digest Press, 1975), 54.
15. Power Reactor Development Co. v. Electricians, 367 U.S. 396 (1961).
16. See John G. Fuller, *We Almost Lost Detroit* (New York: Reader's Digest Press, 1975), for a detailed, if somewhat over-dramatized, description of this accident.
17. Jacquelyn Southern, "Changing Nature: Union Discourse and the Fermi Atomic Power Plant," *International Labor and Working-Class History* 85 (Spring 2014): 44, http://static1.1.sqspcdn.com/static/f/356082/27271503/1475348543780/Changing_Nature_Union_Discourse_and_the.pdf?token=GpFFHEWYTRlCM3fXeecbfzGDpNE%3D.
18. Clarfield and Wiecek, *Nuclear America*, 196.

CHAPTER 4. TO BUILD A NUCLEAR POWER PLANT: ADMIRAL RICKOVER'S WAY

1. Appendix 2 explains the use of a neutron moderator, coolant, and heat transfer in a nuclear reactor.
2. Irving Spiewak and David F. Cope, *Overview Paper on Nuclear Power* (Oak Ridge, TN: Oak Ridge National Laboratory, September 1980), 27, 28.
3. More information on Fermi's reactor can be found in appendix 2.
4. Consumers Power Company, *Final Environmental Statement: Related to the Operation of Midland Plant Units 1 and 2; Docket Nos. 50-329 and 50-330, Consumers Power Company* (Washington, D.C.: U.S. Nuclear Regulatory Commission, July 1982), v.
5. Michael MacRae, "Siegfried Marcus," American Society of Mechanical Engineers, January 6, 2012, https://www.asme.org/engineering-topics/articles/automotive/siegfried-marcus.

6. *Final Environmental Statement*, v.
7. Admiral Hyman G. Rickover, interview by Diane Sawyer, *60 Minutes*, 1984.
8. Willis L. Shirk, "Atoms for Peace in Pennsylvania," *Pennsylvania Heritage* magazine, 35, no. 2, Spring 2009, http://www.phmc.state.pa.us/portal/communities/pa-heritage/atoms-for-peace-pennsylvania.html.
9. Theodore Rockwell, *The Rickover Effect: How One Man Made a Difference* (Annapolis, MD: United States Naval Institute Press, 1992), 20, 21.
10. Paul E. Cantonwine, *The Never-Ending Challenge of Engineering: Admiral H. G. Rickover in His Own Words* (La Grange Park, IL: American Nuclear Society, 2013), xviii.
11. Rockwell, *The Rickover Effect: How One Man Made a Difference*, 21.
12. Rockwell, *The Rickover Effect: How One Man Made a Difference*, 22.
13. Admiral Hyman G. Rickover, interview with Diane Sawyer.
14. Rockwell, *The Rickover Effect: How One Man Made a Difference*, 25.
15. Rockwell, *The Rickover Effect: How One Man Made a Difference*, 23, 25.
16. Rockwell, *The Rickover Effect: How One Man Made a Difference*, 27.
17. Cantonwine, *The Never-Ending Challenge of*, xviii.
18. Shirk, "Atoms for Peace in Pennsylvania."
19. Richard G. Hewlett and Francis Duncan, *Nuclear Navy* (Chicago: University of Chicago Press, 1974), 37.
20. Hewlett and Duncan, *Nuclear Navy*, 78, 90, 91.
21. Rockwell, *The Rickover Effect: How One Man Made a Difference*, 196–197.
22. Shippingport, Pennsylvania, is on the Ohio River about twenty-five miles from Pittsburgh. Rod McCullum, "Reactor Redo," *Scientific American*, May 2019, 73–75. Although zirconium was the best choice available in Rickover's time, if zirconium overheats it can react with water to form hydrogen, which can then explode, as was the case in the Fukushima accident in 2011.
23. Hewlett and Duncan, *Nuclear Navy*, 234.
24. Shirk, "Atoms for Peace in Pennsylvania."
25. Hewlett and Duncan, *Nuclear Navy*, 249.
26. Joint Resolution of Senate and House of Representatives, Public Law 85–826, Resolution to authorize the chairman of the Joint Committee on Atomic Energy to confer a medal on Rear Admiral Hyman George Rickover, United States Navy. August 28, 1958.
27. Joint Committee on Atomic Energy, Hearing on Radiation Safety and Regulation, June 14, 1961, cited in Cantonwine, *The Never-Ending Challenge of Engineering*, 150–151.
28. Cantonwine, *The Never-Ending Challenge of Engineering*, 145.
29. Irvin C. Bupp and Jean-Claude Derian, *Light Water: How the Nuclear Dream Dissolved*

(New York: Basic Books, 1978), 52.

30. Will Davis, "Would Standardized Nuclear Plant Designs Prevent Construction Overruns and Delays? Lessons From the 1970s," *Energy Collective*, March 9, 2015, https://www.energycentral.com/c/ec/would-standardized-nuclear-plant-designs-prevent-construction-overruns-and-delays.

CHAPTER 5. A UTILITY AND A CHEMICAL COMPANY: CAN THIS MARRIAGE ENDURE?

1. "Coal Cars Pile Up Near Dow," *Midland Daily News*, October 28, 1968, 1.
2. "Midland May Be World's First Dual Nuclear Unit," *Midland Daily News*, December 14, 1967, 6.
3. E. N. Brandt, *Growth Company* (East Lansing: Michigan State University Press, 1997), 15.
4. Dorothy Langdon Yates, *Midland, Michigan: The City That Chemistry Built* (Midland: The Piney Woods Group, 2003), 25.
5. George Butterfield, ed., *Bay County, Past and Present* (Bay City, MI: C. & J. Gregory, 1918), 204.
6. Karen Thurlow, ed., *Midland Conservation District Natural Resource Assessment and Strategic Plan 2018* (Midland: Midland Conservation District). Butterfield, *Bay County*, 204.
7. Brandt, *Growth Company*, 15.
8. One of the major markets for bromine was the patent medicine industry. It was a component of Bromo-Seltzer, a patent medicine that took its name from the sodium bromide in its original formula. Jessica D. Griffin, "Bromo-Seltzer, Emerson Drug Co. Baltimore, MD," *Old Main Artifacts* (blog), March 28, 2013, https://oldmainartifacts.wordpress.com/2013/03/28/bromo-seltzer-emerson-drug-co-baltimore-md/.
9. Brandt, *Growth Company*, 19–22.
10. Brandt, *Growth Company*, 21.
11. Gene A. Gillette, *Black Diamond: The Story of Coal in Bay County Michigan* (Bay City, MI: Gene A. Gillette, 2009), 13.
12. George V. Cohee et al., *Coal Resources of Michigan, U.S. Geological Survey, Circular 77* (Washington, D.C.: Department of the Interior, U.S. Geological Survey, 1950), 4–5.
13. J. R. Stark and M. G. McDonald, *Ground Water of Coal Deposits, Bay County, Michigan, Geological Survey of Michigan: Michigan Technological University no. Open-File Report 80-591* (Lansing, MI: U.S. Geological Survey, 1980), 5.
14. Dorothy Langdon Yates, *Salt of the Earth* (Midland: The Midland Historical Society and the Center for the Arts, 1987), 221.

15. "Private Financing Not Enough for N-Plant," *Midland Daily News*, March 23, 1968, 1.
16. Brandt, *Growth Company*, 176–179.
17. Pamela G. Hollie, "The Energy Strategy of the Heaviest Energy User," *New York Times*, February 5, 1978, 145.
18. Hollie, "The Energy Strategy of the Heaviest Energy User," 145.
19. "Dow Midland Employees Live in 2,500 Square-Mile Area," *Midland Daily News*, July 7, 1967, 3A.
20. Marge Messler, discussion with author, October 2017. Messler is a longtime resident of Midland.
21. Gene Goltz, "State Leads A-Power Boom," *Detroit Free Press*, February 2, 1968, 1.
22. Frank Graham Jr., "The Outrageous Mr. Cherry and the Underachieving Nukes," *Audubon*, September 1, 1977, 55.
23. Dr. Robert Asperger, discussion with author, August 18, 2016. Dr. Asperger was employed by the Dow Chemical Company as a chemist in the 1960s.
24. Wolfgang Rudig, *Anti-Nuclear Movements: A World Survey of Opposition to Nuclear Energy* (London: Longman, 1990), 63.
25. Brandt, *Growth Company*, 306.
26. "Company Officials Praise Joint Venture in Midland," *Midland Daily News*, December 14, 1967, 1.
27. "Consumers to Build $267 Million Nuclear Plant for Dow Power Use," *Midland Daily News*, December 14, 1967, 1.
28. "Expanded Dow Job Opportunities Seen," *Midland Daily News*, December 14, 1967, 1.
29. "Milestone in Dow's History, Nuclear Plant Will Help Dow Grow Here," *Brinewell* 26, no. 1, January 1968, 4–6, 4.
30. "Midland May Be World's First Dual Nuclear Unit," 6.
31. "Milestone in Dow's History," 4–6, 4.
32. "Milestone in Dow's History," 4–6, 5.

CHAPTER 6. CONSUMERS POWER: CHALLENGES OF MANAGING NEW NOTIONS

1. Consumers Power Company, *1967 Annual Report* (Jackson, MI: Consumers Power Company, February 16, 1968), 21.
2. George Bush, *Future Builders: The Story of Michigan's Consumers Power Company* (New York: McGraw-Hill Book Company, 1973), 416.
3. "Obituary for James H. Campbell," *Jackson (MI) Citizen Patriot*, January 24, 1972.
4. Bush, *Future Builders*, 403–404. *Technical Information Service: Information Report*

to the Project Companies of the Dow Chemical-Detroit Edison and Associated, Nuclear Power Development Project (Oak Ridge, TN: Technical Information Services, 1953), 14. http://static1.1.sqspcdn.com/static/f/356082/19087025/1341346307347/Nuclear%20 Pow er%20Development%20Project%20-%20Fermi%20I%20Dec%201%201953. pdf?token=jSZ5l7BXCGBES46BnZ%2FITqDH ufA%3D.

5. *Technical Information Service*, 11.
6. Bush, *Future Builders*, 222, 583–584.
7. Bush, *Future Builders*, 429–430.
8. Global Energy Observatory, "Current List of Coal Power Plants," http://globalenergyobservatory.org/list.php?db=PowerPlants&type=Coal.
9. Global Energy Observatory, "Current List of Coal Power Plants."
10. Consumers Power Company, *1973 Annual Report* (Jackson, MI: Consumers Power Company, February 8, 1974), 4.
11. Gene Goltz, "State Leads A-Power Boom," *Detroit Free Press*, January 1, 1968, 1.
12. Consumers Power Company, *1969 Annual Report* (Jackson, MI: Consumers Power Company, February 16, 1970), 9.
13. Consumers Power Company, *1969 Annual Report*, 10.
14. Consumers Power Company, *1969 Annual Report*, 10.
15. John L. Campbell, *Collapse of an Industry* (Ithaca, NY: Cornell University Press, 1988), 95–96. Campbell summarizes the generic benefit for utilities' building of new nuclear generation plants that cost more to build than coal plants in the 1960s and 1970s. Utility accounting is considerably more complicated, but the following brief example does capture the basic economic motivation Consumers Power had to build nuclear plants. In the 1960s, a coal plant with the same generating capacity as the Midland Plant would have cost about $208 million. In 1967 the estimate for building the Midland Plant was $267 million. By spending $59 million more for the nuclear Midland Plant than for a coal plant, Consumers Power would have been able to add an additional $59 million to its rate base. Since the MPSC typically allowed Consumers Power to receive several percent more than their cost of borrowing money to build generation, if Consumers Power was allowed 6 percent more than their cost of borrowing the money to build the Midland Plant, the $59 million more for the Midland Plant versus a coal plant of the same size would mean an additional $3 million more profit for Consumers Power.
16. *Technical Information Service*, 11.
17. Bush, *Future Builders*, 403.
18. Bush, *Future Builders*, 400.
19. Irvin C. Bupp and Jean-Claude Derian, *Light Water: How the Nuclear Dream Dissolved*

(New York: Basic Books, Inc., 1978), 5, 49. A brief explanation of reactor types is provided in appendix 2, and a simplified diagram of the nuclear reactor designed for the Midland Plant is part of appendix 3.

20. Betsy Tompkins, "Big Rock Point: From groundbreaking to greenfield, *Nuclear News*, November 2006, 36, http://www2.ans.org/pubs/magazines/nn/docs/2006-11-3.pdf. In contrast to the 67 MW capacity of the Big Rock Point reactor in the Tompkins article, the NRC page on Big Rock Point lists the capacity as 75 MW ("Big Rock Point," U.S. Nuclear Regulatory Commission, https://www.nrc.gov/info-finder/decommissioning/power-reactor/big-rock-point.html).
21. Bush, *Future Builders*, 545.
22. "Consumers Takes a Gamble with New Nuclear Plant, *Battle Creek (MI) Enquirer*, July 20, 1960, 8.
23. John Scamerhorn, "Consumers' First Nuclear Generation Plant Begun," *Battle Creek (MI) Enquirer*, July 24, 1960, section 2, 9.
24. Richard G. Hewlett and Francis Duncan, *Nuclear Navy* (Chicago: University of Chicago Press, 1974), 255.
25. Glenn T. Seaborg, *Civilian Nuclear Power: A Report to the President, 1962* (Washington, D.C.: U.S. Atomic Energy Commission, November 20, 1962), 34.
26. "JCP & L Plans Nuclear Generator Plant," *Central New Jersey Home News* (New Brunswick), December 12, 1963, 21.
27. Robert Perry, *Development and Commercialization of the Light Water Reactor, 1946–1976* (Santa Monica, CA: Rand Corporation, 1976), 35. http://large.stanford.edu/courses/2015/ph240/morganl/docs/R2180.pdf.
28. Gerard H. Clarfield and William M. Wiecek, *Nuclear America: Military and Civilian Nuclear Power in the United States, 1940–1980* (New York: Harper & Row, 1984), 278.
29. "Atom Plants Mushrooming," *Detroit Free Press*, July 23, 1967, 12a.
30. "$100 Million Electrical Plant Set For Van Buren," *News-Palladium* (Benton Harbor, MI), January 28, 1966, Final Edition 1, 9.
31. "A Turn to Bits & Pieces for Nuclear," *Electric World* 165, no. 16, April 18, 1966, 62.
32. Perry, *Development and Commercialization of the Light Water Reactor*, 29, 37.
33. "$100 Million Electrical Plant Set For Van Buren," *News-Palladium* (Benton Harbor, MI), 1, 9.
34. Allan Sloan, "Consumers: A Sorry Tale of a Short-Circuited Utility," *Detroit Free Press*, June 20, 1974, 5.
35. Gina Shaw, *Building for the Future: The Story of Consumers Energy, 1886–2016* (St. Louis, MO: Essex Publishing Group, Inc., 2017) 54.

36. "Consumers Power Sues Nuclear Plant Builders," *Lansing State Journal*, August 29, 1974, 3.
37. Charles Komanoff, *Power Plant Cost Escalation* (New York: Van Nostrand Reinhold Company, 1981), 298, 301.
38. "Consumers Eying Nuclear Plant on Saginaw Bay," *Detroit Free Press*, August 4, 1971, 45.
39. "Consumers Buys 2 Big A-Generators," *Detroit Free Press*, December 9, 1972, 21. "Consumers Seeks Okay for Saginaw Nuclear Plant," *Holland (MI) Evening Sentinel*, November 1, 1973, 7.
40. "Consumers Seeks Okay for Saginaw Nuclear Plant," 7.
41. Bush, *Future Builders*, 418.
42. "Consumers Eying Nuclear Plant on Saginaw Bay," 45.
43. Barbara Stanton, "Can It Start Producing Power, Not Problems?," *Detroit Free Press*, July 27, 1968, 17.
44. "Fermi, Unit 2," U.S. Nuclear Regulatory Commission, https://www.nrc.gov/info-finder/reactors/ferm2.html.
45. "Obituary for James H. Campbell."
46. Bush, *Future Builders*, 449.
47. Bush, *Future Builders*, 417–418.

CHAPTER 7. LET BECHTEL DO IT: JUST MAKE IT A MEGAPROJECT

1. Richard D. James, "Bechtel: Industrial Giant," *Daily Item* (Sunbury, PA), April 11, 1979, 1.
2. James, "Bechtel: Industrial Giant," 3.
3. James, "Bechtel: Industrial Giant," 1.
4. David Tomlinson, "Fidelity Fiduciary Bank," *Mary Poppins*, directed by Robert Stevenson, music and lyrics by Robert B. Sherman and Richard M. Sherman (Burbank, CA: Walt Disney, 1964).
5. "Study by Bechtel Preceded Planning for Nuclear Plant," *Midland Daily News*, December 14, 1967, 6.
6. "Study by Bechtel Preceded Planning for Nuclear Plant," 6.
7. Guy Snyder, "Nuclear Power in Michigan," *Michigan Contractor & Builder* (Detroit), July 24, 1976, 9.
8. "History," Bechtel, https://www.bechtel.com/about-us/history/.
9. Bonnie Holley, "Steve Bechtel Jr. Visits Midland Project," *Midland Reactor*, July 1, 1981, 1. When Steve Bechtel Jr. made his visit to the Midland Plant in 1981, the estimated cost to complete the Midland plant had increased from the original 1967 estimate of $267 million to $3.1 billion (details of the cost increases are given in chapter 21).

10. Wives' shares were held in trust by their husbands, who retained voting power. Mark Dowie, "The Bechtel File," *Mother Jones*, SEPT / OCT 1978, 33. In an article in *Fortune*, Shawn Tully wrote that in 2016 Bechtel was owned by fifty shareholders and that 40 percent of the shares were owned by the Bechtel family members. Shawn Tully, "Meet the Private Company That Has Changed the Face of the World," *Fortune*, May 17, 2016.
11. Dowie, "The Bechtel File," 33.
12. Sally Denton, *The Profiteers, Bechtel and the Men Who Built the World* (New York: Simon & Schuster, 2016), 53, 54, 74.
13. Ronald Brownstein and Nina Easton, *Reagan's Ruling Class: Portraits of the President's Top One Hundred Officials* (New York: Pantheon Books, 1983), 151. Denton, *The Profiteers, Bechtel and the Men Who Built the World*, 71.
14. Denton, *The Profiteers, Bechtel and the Men Who Built the World*, 71.
15. "Study by Bechtel Preceded Planning for Nuclear Plant," 1.
16. "Giant A-Plant Will Go Up in Dade," *Miami Herald*, November 16, 1965, 1.
17. Telephone interview on December 15, 2016, with engineer who worked in Bechtel's Ann Arbor office in the 1970s. Name is withheld at his request.
18. Denton, *The Profiteers, Bechtel and the Men Who Built the World*, 49.
19. "A Turn to Bits & Pieces for Nuclear," *Electric World* 165, no. 16, April 18, 1966, 62.
20. Affidavit provided to me from the estate of Sara Hughes (1933–2018).
21. Patrick Fitzgerald, "Bechtel Thrives on Midland Plant," *Lansing State Journal*, April 12, 1984, 3B.

CHAPTER 8. 1968: FORCES NOT UNDER YOUR CONTROL MATTER

1. "C-C Told of New N-Plant," *Midland Daily News*, January 9, 1968, 1.
2. Geoffrey C. Ward and Ken Burns, *The Vietnam War* (New York: Alfred A. Knopf, 2017), 160.
3. Ward and Burns, *The Vietnam War*, 160.
4. Ward and Burns, *The Vietnam War*, 98.
5. Mark Lorell and Charles Kelley Jr., *Casualties, Public Opinion, and Presidential Policy During the Vietnam War* (Santa Monica, CA: The Rand Corporation, March 1985), 19. https://www.rand.org/content/dam/rand/pubs/reports/2007/R3060.pdf.
6. Louis H. Woods, "Protesters Fail to Budge Dow: Napalm Production Will Go On," *Midland Daily News*, May 9, 1968, 1.
7. Edwin L. Dale Jr., "What Vietnam Did to the American Economy," *New York Times*, January 28, 1973, 141–142. This article provides a good review of how President Johnson's failure to recommend higher taxes in early 1968 led to price inflation in the late 1960s.

8. Tom Riddell, "Inflationary Impact of the Vietnam War," *Vietnam Generation* 1, no. 1, Article 4 (1989), 57–58, http://digitalcommons.lasalle.edu/cgi/viewcontent.cgi?article=1003&context=vietnamgeneration.
9. "Consumer Price Index, 1913–," Federal Reserve Bank of Minneapolis, https://www.minneapolisfed.org/community/financial-and-economic-education/cpi-calculator-information/consumer-price-index-and-inflation-rates-1913.
10. Consumers Power Company, *1967 Annual Report* (Jackson, MI: Consumers Power Company, February 16, 1968), 3. Consumers Power Company, *1970 Annual Report* (Jackson, MI: Consumers Power Company, February 15, 1971), 12.
11. U.S. Census, "1189. Price and Cost Indexes for Construction," Allcountries.org,https://allcountries.org/uscensus/1189_price_and_cost_indexes_for_construction.html.
12. Betty Herring, "Nuclear plant Called 'Greatest Investment in History of State," *Midland Daily News*, June 27, 1969, 1.
13. Consumers Power Company, *1968 Annual Report* (Jackson, MI: Consumers Power Company, February 14, 1969), 12.
14. "Midlander Wounded in Vietnam," *Midland Daily News*, May 29, 1968, 3.
15. Consumers Taps Ohio Firm for N-steam Supply Systems," *Midland Daily News* May 29, 1968, 3.

CHAPTER 9. MARY SINCLAIR: NO TYPICAL 1960S WOMAN

1. Sandra Collinson, letter to the editor, *Midland Daily News*, October 10, 1980, 6.
2. Mary Sinclair, letter to the editor, *Midland Daily News*, June 25, 1969.
3. Mary Sinclair, *Nuclear Power & Public Concern*, 1968 and revised 1971, Mary P. Sinclair papers 1967–1999, Bentley Historical Library, University of Michigan.
4. "Midland N-Plant Project Debated," *Midland Daily News*, December 19, 1969, 1.
5. "Nuclear Power of Importance Here," editorial, *Midland Daily News*, December 20, 1969, 10.
6. Mary P. Sinclair, "Letter to the Editor," *Midland Daily News*, December 23, 1969, 8.
7. John Catenacci, "Mary Sinclair Talks about Midland and the Price of Speaking Out," *Tricity Magazine* (Midland), 1976, 8. This magazine is available at the Grace A. Dow Memorial Library in Midland, Michigan.
8. *Sixty Minutes*, "Mary and Goliath," produced by Christine Weicher, aired January 27, 1985, Columbia Broadcasting System. To view this recording, see "60 Minutes: Midland Nuclear Hearing, Part 1," YouTube video, https://www.youtube.com/watch?v=cbQso2PC6Jc; and "60 Minutes: Midland Nuclear Hearing, Part 2," YouTube video, https://www.youtube.

com/watch?v=Tn_PiPHBXMg.

9. "Man's Social Progress Has Fallen Far Behind, Sinclair Tells Seniors," *Midland Daily News*, June 13, 1968, 1.
10. Frank Graham Jr., "The Outrageous Mr. Cherry and the Underachieving Nukes," *Audubon*, September 1977, 58.
11. Peter Sinclair, discussion with author, November 2017.
12. Obituary for Mary Sinclair, *Midland Daily News*, January 15, 2011. Catenacci, "Mary Sinclair Talks about Midland," 8.
13. Terry Miller, discussion with author, January 2017.
14. Ray H. Boundy and Raymond F. Boyer, *Styrene: Its Polymers, Copolymers, and Derivatives, American Chemical Society, Monograph Series* (New York: Reinhold Publishing Corp., 1952), ix.
15. Graham, "The Outrageous Mr. Cherry and the Underachieving Nukes," 55.
16. Graham, "The Outrageous Mr. Cherry and the Underachieving Nukes," 55.
17. G. M. Woodwell, "Toxic Substances an Ecological Cycles," *Scientific American* 216, no. 3, March 1, 1967, 24–31.
18. Catenacci, "Mary Sinclair Talks about Midland," 9.
19. Catenacci, "Mary Sinclair Talks about Midland," 10. See also Mary P. Sinclair, *Nuclear Power & Public Concern: Statements and Questions*, Mary P. Sinclair papers, 1967–1999, Bentley Historical Library, University of Michigan, call number DA 2 S6165 N964. Excerpts from this booklet can be found in appendix 1. Dorothy Dow Arbury's support of Mary Sinclair is also documented in Graham, "The Outrageous Mr. Cherry and the Underachieving Nukes," 55.
20. Sally Moss, discussion with author December 29, 2016.
21. Catenacci, "Mary Sinclair Talks about Midland," 11. In January 2018, Gary Skory, who was the director of the Midland County Historical Society, said that "City of Beautiful Churches" was a term the Midland Chamber of Commerce coined in the 1950s to describe Midland.
22. Catenacci, "Mary Sinclair Talks about Midland," 11.
23. Mary Sinclair to Liza Porter Trojan, "Sympathy letter on the death of Dr. William Porter, father of Liza Porter Trojan," May 12, 1999, Mary P. Sinclair papers, 1967–1999, Bentley Historical Library, University of Michigan. Sinclair received a master's degree in environmental communications from the School of Natural Resources at the University of Michigan in May 1973. The title of her thesis was "Nuclear Environmentalism—A Case Study."
24. "Two Urge Acceleration of Fusion Power Reactors," *Evening Sun* (Hanover, PA), May 20,

1971, 20.

25. Mary Sinclair, letter to the editor, *Wisconsin State Journal* (Madison), August 23, 1974, 8.
26. Mary P. Sinclair, "Public Information Provided on Nuclear Power," Mary P. Sinclair papers, 1967–1999, Bentley Historical Library, University of Michigan.
27. Mary P. Sinclair, résumé, Mary P. Sinclair papers, 1967–1999, Bentley Historical Library, University of Michigan. John Sherwood, "Foes Ask Hearings Be Ruled Illegal," *Battle Creek (MI) Enquirer*, June 25, 1970, 12.
28. "GOP Club to Hear Nuclear Discussion," *Midland Daily News*, November 12, 1970, 3.
29. Constance E. Cook, *Nuclear Power and Legal Advocacy* (Lexington, MA: Heath and Company, 1980), 51–53. A separate group consisting of eight homeowners and business men called the Mapleton Intervenors also participated in the licensing hearings. Led by Wendell Marshall, the Mapleton Intervenors were mostly concerned with the effect the Midland Plant, which was located just upstream on the Tittabawassee from Mapleton, could have on their property values. Mary Sinclair chose not to join with the Mapleton Intervenors because she saw their interests as local in nature.
30. Cook, *Nuclear Power and Legal Advocacy*, 52.
31. Dr. Robert Asperger, discussion with author, August 18, 2016.
32. John Emshwiller, "Nuclear Nemesis: Using the Law's Delay, Myron Cherry Attacks Atomic-Power Projects," *Wall Street Journal*, March 10, 1978, 1.
33. Christoph Hohenemser, Roger Kasperson, and Robert Kates, "The Distrust of Nuclear Power," *Science* 196, April 1, 1977, 27, https://www.ncbi.nlm.nih.gov/pubmed/841337.
34. Gay McGee, "Midland Housewife Initiated Nuclear Plant Hassle," *Bay City (MI) Times*, June 6, 1971, C1.
35. Dr. Connie Moll, email to author, September 24, 2019.
36. Mary P. Sinclair, "Great Lakes Commentary: The Nuclear Threat To The Great Lakes," *Toledo Journal of Great Lakes' Law, Science & Policy* 2, no. 33 (Fall 1999): 34.

CHAPTER 10. HOW TO REACT TO MIDLAND PLANT CRITICS: SHAME, ORGANIZE, LABEL

1. C. E. Cook, *Nuclear Power and Legal Advocacy* (Lexington, MA: Heath and Company, 1980), 59.
2. "Committee Seeks N-plant Approval," *Midland Daily News*, October 16 1970, 1.
3. "Private Financing Not Enough for N-Plants," *Midland Daily News*, March 13, 1968, 1.
4. "Dow Says Jobs Hang on Nuclear Power," *Lansing State Journal*, February 4, 1977, 10.
5. "Dr. Wayne North," obituary, *Daily Telegram* (Adrian, MI), September 19, 1989, 2.
6. Robert A. Richardson, "The Selling of the Atom," *Bulletin of the Atomic Scientist* 30, no. 8

(October 10, 1974): 28. Oak Ridge Associated Universities, *25th Annual Report for the Year Ending June 30, 1971* (Oak Ridge, TN: Oak Ridge Associated Universities, 1971), 38. In 1974 the Atomic Energy Commission (AEC) was divided into two agencies: the U.S. Energy Research and Development Administration and the U.S. Nuclear Regulatory Commission (NRC). The NRC was given the licensing, rulemaking, and safety functions that had been done by the AEC. A reference on the history of nuclear regulation is J. Samuel Walter and Thomas R. Wellock, "A Short History of Nuclear Regulation, 1946–2009," Office of the Secretary, U.S. Nuclear Regulatory Commission, NUREG/BR-0175, October 2010, http://www.nrc.gov/docs/ML1029/ML102980443.pdf.

7. Oak Ridge Associated Universities, *25th Annual Report for the Year Ending June 30, 1971*, 1, 5.
8. Jerry M. Flint, "Midland Mich., Wants Nuclear Power," *New York Times*, December 3, 1971, 24.
9. Bud Farrar, "Dow Project Challenged in Company Town," *Edwardsville (IL) Intelligencer*, April 20, 1974, 2.
10. Farrar, "Dow Project Challenged in Company Town," 2. Clearly Dr. North did not know how to cite odds. I am sure he meant to say one in a billion.
11. U.S. Atomic Energy Commission, "Reactor Safety Study—An Assessment of Accident Risks in U.S. Commercial Power Plants" (WASH-1400/NRG) (Rasmussen Report), October 1, 1975, 3. Critics of WASH-1400 and earlier estimates of the risk of nuclear power claimed that all of these estimates drastically underestimated the risk of nuclear power. For two examples of these critiques see Gerald H. Clarfield and William M. Wiecek, *Nuclear America: Military and Civilian Nuclear Power in the United States, 1940–1980* (New York: Harper & Row, Publishers), 1984, 374–375; and Irvin C. Bupp and Jean Claude Derian, *Light Water: How the Nuclear Dream Dissolved* (New York: Basic Books, Inc., Publishers), 1978, 122–123.
12. Richardson, "The Selling of the Atom," 29, 30. Oak Ridge Associated Universities, *25th Annual Report for the Year Ending June 30, 1971*, 34.
13. Richardson, "The Selling of the Atom," 30.
14. "Midlanders Gain N-plant Answers at Oak Ridge," *Midland Daily News*, November 11, 1970, 3.
15. Richardson, "The Selling of the Atom," 30.
16. "Nuclear Power Committee Completes Nine Workshops," *Midland Daily News*, November 28, 1970, 3.
17. Richardson, "The Selling of the Atom," 30.
18. *Sixty Minutes*, "Mary and Goliath," produced by Christine Weicher, aired January 27, 1985,

Columbia Broadcasting System, "60 Minutes: Midland Nuclear Hearing, Part 2," YouTube video, https://www.youtube.com/watch?v=Tn_PiPHBXMg.

19. "N-Plant Backers Refute Claims by Mrs. Sinclair," *Midland Daily News*, August 23, 1971, 1.
20. Farrar, "Dow Project Challenged in Company Town," 2.
21. "N-Plant Construction Delayed Pending OK," *Midland Daily News*, 14/11 1970, 1.
22. Casey Bukro, "Conservationists Urge AEC to Set Antipollution Standards," *Chicago Tribune*, July 22, 1970, 19.
23. Yvonne Zipp, "Palisades Nuclear Plant: A Timeline of Events," *MLive* (Kalamazoo, MI), August 24, 2014, https://www.mlive.com/news/kalamazoo/2014/08/palisades_nuclear_plant_timeli.html. Jim Donahue, "Pre-Hearings Talks Held in Covert," *News-Palladium* (Benton Harbor, MI), February 11, 1967, 10. Consumers Power Company, *1967 Annual Report* (Jackson, MI: Consumers Power Company, February 16, 1968), 14.
24. "Atomic Plant Hearing Begins," *Ironwood (MI) Daily Globe*, June 23, 1970, 3.
25. Consumers Power Company, *1971 Annual Report* (Jackson, MI: Consumers Power Company, February 16, 1972), 7.
26. "Nuclear Power Will Help Dow Grow Here," *Brinewell*, January 1986, 5, 6.
27. Flint, "Midland Mich., Wants Nuclear Power," 24.
28. Alan Ott provided me a copy of this memorandum to me in 2018.
29. "Midland N-Plant Given Priority Status," *Midland Daily News*, October 28, 1971, 1.
30. Claudia Protz, "December Start-up of N-hearing Hinted," *Midland Daily News*, 1.
31. Cook, *Nuclear Power and Legal Advocacy*, 60.
32. Linda Lein, "Project Gains Momentum," *Midland Reactor*, November 1974, 4.
33. Cook, *Nuclear Power and Legal Advocacy*, 61.

CHAPTER 11. MIDLAND PRIDE: WHY MIDLANDERS SUPPORTED THE MIDLAND NUCLEAR PLANT

1. "Structures," Mid-Century Modern Midland, https://midcenturymidland.org/structures/.
2. E. N. Brandt, *Growth Company* (East Lansing: Michigan State University Press, 1997), 70.
3. E. N. Brandt, *Chairman of the Board* (East Lansing: Michigan State University Press, 2003), 186.
4. Brandt, *Growth Company*, 268–270.
5. Dorothy Langdon Yates, *Salt of the Earth* (Midland: The Midland Historical Society and the Center for the Arts, 1987), 162–163.
6. Yates, *Salt of the Earth*, 243–244. This book lists only a few of the foundation gifts and gifts made by descendants of Dow Chemical pioneers to projects in Midland. Some of

the other foundations are Rollin M. Gerstacker, Charles J. Strostacker, Elsa U. Pardee, Macauley and Helen Dow Whiting, Allen, Barstow, Herbert D. and Junia Doan, and Herbert H. and Barbara C. Dow. For an appreciation of amounts and scope of gifts made to the Midland County see Yates, *Salt of the Earth*, 246–247.

7. Brandt, *Chairman of the Board*, 186.
8. "Midland Clergy Support Dow," *Midland Daily News*, May 9, 1968, 1.
9. "Midland Clergy Support Dow," 1.
10. Robert M. Neer, *Napalm: An American Biography* (Cambridge, MA: Harvard University Press, 2013), 6, 84, 225, 299.
11. Curtis E. LeMay with MacKinlay Kantor, *Mission with LeMay: My Story* (Garden City, NY: Doubleday & Company, Inc., 1965), 387. Napalm is the ideal incendiary weapon: cheap and stable, and a burning gel that sticks to roofs, furniture, and skin. The big jump in napalm use for Vietnam paralleled the development of a new, hotter compound called napalm B—the mixture of 46 percent crystalline polystyrene, 21 percent benzene, and 33 percent gasoline. Bob Horton, "Use of Napalm Unprecedented in History of Warfare," *Midland Daily News*, March 19, 1968, 7. Napalm B was developed in 1964 by scientists at Eglin Air Force Base in Florida. It burned hotter, stuck tighter, and ignited more reliably than earlier formulations. Neer, *Naplam*, 116.
12. This figure did not include the quantity the Navy used in bombs or the army's use in flamethrowers. Bob Horton, "Use of Napalm Unprecedented in History of Warfare," *Midland Daily News*, March 19, 1968, 7.
13. Bob Horton, "Use of Napalm Unprecedented in History of Warfare," 7.
14. "Midland Clergy Support Dow," 1.
15. "Demonstrators Protest Dow's Making of Napalm," *Ironwood (MI) Daily Globe*, May 8, 1968, 1.
16. Kevin Roose, "Why Napalm Is a Cautionary Tale For Tech Giants," *New York Times*, March 6, 2019, B1, B5, https://www.nytimes.com/2019/03/04/technology/technology-military-contracts.html.
17. Bridge Magazine and The Detroit Journalism Cooperative, *The Intersection: What Detroit Has Gained, and Lost, 50 Years After the Uprising of 1967* (Traverse City, MI: Mission Point Press, 2017), xvii, xviii, 28.
18. EllaMarie Schroeder and H. J. Smith, "Voices: Extraordinary Women of Midland County, Phae Dorman," *The Midland Log: A Journal of the Midland County Historical Society of Midland County Center for the Arts* (Fall/Winter 2011), 13.
19. Schroeder and Smith, "Voices," 41.
20. *Extraordinary Women of Midland County*, exhibit at the Doan History Center by the

Midland Center for the Arts and the Midland County Historical Society, 2011–2012, Tina S. Van Dam, Chair of Steering Committee. The cross that was burned in Dorothy Fauver's front lawn was part of the 2011–2012 exhibit, but it unfortunately appears to have been lost when the Tittabawassee River flooded Midland in 2020.

CHAPTER 12. MYRON CHERRY: BRIGHT, DEDICATED, AND MEAN

1. Myron Cherry, telephone discussion with author, March 5, 2017.
2. Frank Graham Jr., "The Outrageous Mr. Cherry and the Underachieving Nukes," *Audubon*, September 1977, 56.
3. Cherry, telephone discussion.
4. John R. Emshwiller, "Nuclear Nemesis: Using the Law's Delay, Myron Cherry Attacks Atomic-Power Projects," *Wall Street Journal*, March 10, 1978, 1.
5. Consumers Power, *1971 Annual Report* (Jackson, MI: Consumers Power Company, February 16, 1972), 7.
6. Emshwiller, "Nuclear Nemesis," 1.
7. Emshwiller, "Nuclear Nemesis," 1.
8. C. E. Cook, *Nuclear Power and Legal Advocacy* (Lexington, MA: Heath and Company, 1980), 52. Cherry's anti-nuclear activities were not limited to Palisades Plant and the Midland Plant. With his victory at the Palisades Plant, he gained a national representation. He was the chief attorney for the Consolidated National Intervenors at the emergency core cooling systems rulemaking hearings and represented clients in about two dozen licensing battles from Rhode Island to Wisconsin.
9. Cook, *Nuclear Power and Legal Advocacy*, 53.
10. Cook, *Nuclear Power and Legal Advocacy*, 55.
11. Irving Like, "Multi-Media Confrontation—The Environmentalists' Strategy for a No-Win Agency Proceeding," *Ecology Law Review* 1, no. 3 (1971): 495.
12. "N-plant Opponents Slow AEC Hearing," *Midland Daily News*, December 1, 1970, 1."Nuclear Power Plant Meeting Opposition." *Berkshire Eagle* (Pittsfield, MA), May 20, 1969, 18.
13. "Minnesota Nuclear Plant Generates Much Debate," *Asbury (NJ) Park Press*, September 7, 1969, 22.
14. Gordon Hines, telephone conversation with author, January 27, 2017.
15. Graham, "The Outrageous Mr. Cherry and the Underachieving Nukes," 57.
16. Peter Sinclair, discussion with author, November 2, 2017.
17. Graham, "The Outrageous Mr. Cherry and the Underachieving Nukes," 57.

18. Dr. Robert G. Asperger, letter to Moody's and Standard & Poor's Bond Rating Departments, January 6, 1976, Mary Sinclair's papers, 1967–1999, Bentley Historical Library, University of Michigan, Ann Arbor, Michigan.
19. Dr. Robert G. Asperger, telephone interview with author, August 18, 2016.
20. Graham, "The Outrageous Mr. Cherry and the Underachieving Nukes," 57.
21. Claudia Protz, "Cherry Quizzes 3 Witnesses on Nuclear Evacuation Plan," *Midland Daily News*, July 13, 1971, 1.
22. Graham, "The Outrageous Mr. Cherry and the Underachieving Nukes," 57.
23. Emshwiller, "Nuclear Nemesis," 1.
24. Emshwiller, "Nuclear Nemesis," 1.
25. Cook, *Nuclear Power and Legal Advocacy*, 55.
26. Calvert Cliffs' Coordinating Committee, Inc., et al., Petitioners, v. United States Atomic Energy Commission and United States of America, Respondents, Baltimore Gas and Electric Company, Intervenor. Calvert Cliffs' Coordinating Committee, Inc., et al. Petitioners, v. United States Atomic Energy Commission and United States of America, Respondents, 449F.2d 1109 (D.C. Cir. 1971), https://law.resource.org/pub/us/case/reporter/F2/449/449.F2d.1109.24839.24871.html.
27. "Midland Nuclear Plant Delay: Whose Fault?," *Bay City (MI) Times*, December 23, 1972, 4-A.
28. J. Samuel Walker, *Three Mile Island: A Nuclear Crisis in Historical Prospective* (Berkeley: University of California Press, 2004), 54–60.
29. "All Parties Agree to Resume Hearing," *Midland Daily News*, June 8, 1971, 1.
30. "Midland Nuclear Plant Delay," 4-A.
31. *Sixty Minutes*, "Mary and Goliath," produced by Christine Weicher, aired January 27, 1985, Columbia Broadcasting System, "60 Minutes: Midland Nuclear Hearing, Part 2," YouTube video, https://www.youtube.com/watch?v=Tn_PiPHBXMg.
32. "National Environmental Policy Act," Administrative Conference of the United States and the American Bar Association's Section of Administrative Law and Regulatory Practice, December 21, 2019, https://sourcebook.acus.gov/wiki/National_Environmental_Policy_Act/view.
33. Calvert Cliffs' Coordinating Committee, Inc. et al, Petitioners v. United States of America, Respondents Decision, Selected materials on the Calvert Cliffs' Decision, Its Origin and Aftermath, Joint Committee on Atomic Energy, Congress of the United States, 92d Congress, 1st session, February 1972, 127. https://www.google.com/books/edition/Selected_Materials_on_the_Calvert_Cliffs/0MSljC0anl8C?hl=en&gbpv=1.
34. Cook, *Nuclear Power and Legal Advocacy*, 73–74.

35. Environmental Defense Fund, Incorporated et al., Petitioners, v. William D. Ruckelshaus, Administrator of the Environmental Protection Agency & Environmental Protection Agency, Respondents, Izaak Walton League of America, Montrose Chemical Corporation of California, State of New York, Intervenors. 439 F.2d 584 (1971), https://scholar.google.com/scholar_case?case=9832312075337683963&q=environmental+defense+fund+v.+ruckelshaus&hl=en&as_sdt=80000006&as_vis=1.
36. Gillian Metzger, "The Story of Vermont Yankee: A Cautionary Tale," Paper Number 05-92, *Columbia Law School*, June 14, 2005, 26.
37. The court decision in the Midland case with regard to radioactive waste was controlled by the Vermont Yankee Case, *Natural Resources Defense Council v. U.S. Nuclear Regulatory Commission*, which was handed down on the same day. Nelson Aeschliman et al., Petitioners, v. U.S. Nuclear Regulatory Commission and United States of America, Respondents, Consumers Power Co., a Michigan Corp., Intervenor. Saginaw Valley Nuclear Study Group et al., Petitioners, v. U.S. Nuclear Regulatory Commission and United States of America, Respondents, Consumers Power Company, Intervenor, 547 F.2d 622 (D.C. Cir. 1976), 31, 24, 40, 39, Open Jurist, https://openjurist.org/547/f2d/622/aeschliman-v-united-states-nuclear-regulatory-commission. The provision in the opinion to reevaluate the changed circumstances regarding Dow Chemical's need for process steam was based on the fact that when Dow Chemical and Consumers Power renegotiated their contract in February of 1974, and the renegotiated contract allowed Dow Chemical to retain its coal-fired plant. Natural Resources Defense Council, Inc., et al., Petitioners, v. U.S. Nuclear Regulatory Commission and United States of America, Respondents, Vermont Yankee Nuclear Power Corporation, Intervenor. Natural Resources Defense Council, Inc., and Consolidated National Intervenors, Petitioners v. U.S. Nuclear Regulatory Commission and United States of America, Respondents, Baltimore Gas and Electric Co. et al., Intervenors. 547 F.2d 633 (D.C. Cir. 1976), Open Jurist, https://openjurist.org/547/f2d/633/natural-resources-defense-council-inc-v-united-statesnuclear-regulatory-commission.
38. Vermont Yankee Nuclear Power Corporation v. Natural Resources Defense Council, Inc., and Consumers Power Company v. Aeschliman, 435 U.S. 519 (1978); 98 S. Ct. 1197 (1978).
39. Vermont Yankee Nuclear Power Corporation v. Natural Resources Defense Council, Inc., and Consumers Power Company v. Aeschliman, 435 U.S. 519 (1978); 98 S. Ct. 1197 (1978), 556–557.
40. Vermont Yankee Nuclear Power Corporation v. Natural Resources Defense Council, Inc., and Consumers Power Company v. Aeschliman, 549. http://large.stanford.edu/courses/2017/ph241/dadabbo2/docs/vermont-yankee.pdf.

41. Vermont Yankee Nuclear Power Corporation v. Natural Resources Defense Council, Inc., and Consumers Power Company v. Aeschliman, 28, fn. 22.
42. Vermont Yankee Nuclear Power Corporation v. Natural Resources Defense Council, Inc., and Consumers Power Company v. Aeschliman, 539. James F. Raymond, "A *Vermont Yankee* in King Burger's Court: Constraints on Judicial Review Under NEPA," *Boston College Environmental Affairs Law Review* 7, no. 4August 1, 1979, 632–633, https://lawdigitalcommons.bc.edu/cgi/viewcontent.cgi?referer=https://scholar.google.com/&httpsredir=1&article=1825&context=ealr.
43. Vermont Yankee Nuclear Power Corporation v. Natural Resources Defense Council, Inc., and Consumers Power Company v. Aeschliman, 557, 558.
44. "Nuclear Fuel," Nuclear Energy Institute, Washington, D.C., https://www.nei.org/fundamentals/nuclear-fuel.
45. Eric Sharp, "Consumers Power Shows of A-plant," *Detroit Free Press*, April 21, 1979, 40.
46. Mitch Jacoby, "As Nuclear Waste Piles Up, Scientists Seek the Best Long-Term Solutions," *Chemical & Engineering News* 98, no. 12, March 30, 2020, https://cen.acs.org/environment/pollution/nuclear-waste-pilesscientists-seek-best/98/i12.
47. "Interview with Mary Sinclair," *Day by Day*, WUCM-TV, October 2, 1976. Mary P. Sinclair papers, 1967–1999, Bentley Historical Library, University of Michigan.
48. Emshwiller, "Nuclear Nemesis," 1.

CHAPTER 13. CONSTRUCTION QUALITY CONTROL: AN UNREDEEMED PROMISE OF REFORMATION

1. Linda Lein, "Project Gains Momentum," *Midland Reactor*, November 1, 1976, 4.
2. Lein, "Project Gains Momentum," 4.
3. Midland Environmental Report, Operating License Stage, Volume 2, Revision 1, April 1978, 2.5-2 and 2.5-3, Midland County Historical Society Document.
4. Bruce Peck, discussion with author, April 8, 2016.
5. "Construction Halted on Nuclear Plant," *Battle Creek (MI) Enquirer*, November 15, 1970, 2.
6. Lein, "Project Gains Momentum," 4.
7. "Consumers to Await Construction Permit," *Lansing State Journal*, November 16, 1970, 4.
8. Peck, discussion with author.
9. David Everett, "Decade of Mistakes at Midland A-Plant," *Detroit Free Press*, November 30, 1981, 13.
10. Aeschliman v. Nuclear Regulatory Commission, No. 547 F .2d 622, 625 (D.C. Cir. 1976), fn. 3, https://law.resource.org/pub/us/case/reporter/F2/547/547.F2d.622.73-1867.73-1776.

html.

11. Bud Farrar, "Midland Nuclear Debate Costly," *Decatur (IL) Daily Review*, April 16, 1974, 8.
12. Linda Lein, "Cadwelding Continues in Containments," *Midland Reactor*, March 1, 1977, 1.
13. John Emshwiller, "How a Thriving Utility Became a Sagging One After Its Giant Step," *Wall Street Journal*, October 8, 1974, 1.
14. Everett, "Decade of Mistakes at Midland A-Plant," 13.
15. Joy E. Buchanan, "Nuclear Plant Foes Won't Ask Hearing," *Midland Daily News*, December 22, 1973, 1.
16. "Michigan A-Plant Welding Ban Lifted," *Chicago Tribune*, December 19, 1973, 7.
17. David Stoll, "Midland Nuke Plant Radiates Scandal," *Ann Arbor Sun*, December 20, 1974, 16, https://aadl.org/node/198689.
18. Saul Friedman, "Order Halting Nuclear Plant Is Reversed," *Philadelphia Inquirer*, December 21, 1973, 5.
19. Allan Sloan, "Utilities Nuclear Nightmare," *Detroit Free Press*, February 17, 1974, 24.
20. Sloan, "Utilities Nuclear Nightmare, 4.
21. Bruce Peck, discussion with author, April 20, 2016.
22. Bob Wheeler, discussion with author, May 24, 2016.
23. "Are Nuclear Plants Safe? Midland Is No Assurance," editorial, *Detroit Free Press*, July 30, 1974, 8.
24. "Atomic Safety and Licensing Board, Consumers Power Company (Midland Plant, Units 1 and 2), 15 NRC 1060 (1982), April 30, 1982," in *Nuclear Regulatory Commission Issuances: Opinions and Decisions of the Nuclear Regulatory Commission with Selected Orders, January 1, 1982–April 30, 1982*, vol. 15, book 1 (Washington, D.C.: U.S. Nuclear Regulatory Commission, 1982), 1066.
25. "Are Nuclear Plants Safe?," 8.

CHAPTER 14. 1975: CARRY ON WITH THE MIDLAND PLANT

1. Consumers Power Company, *1974 Annual Report* (Jackson, MI: Consumers Power Company, February 11, 1975), 2.
2. Bernard Reich, *Securing the Covenant: United States-Israel Relations after the Cold War* (Westport, CT: Greenwood Publishing Group, Inc., 1995), cited in Michael Corbett, "Oil Shock of 1973–74," Federal Reserve History, November 23, 2013, https://www.federalreservehistory.org/essays/oil_shock_of_1973_74.
3. Corbett, "Oil Shock of 1973–74.
4. "OPEC Oil Price Annually 1960–2019," Statista, Business Data Platform, https://www.

statista.com/statistics/262858/change-in-opec-crude-oil-prices-since-1960/.

5. William K. Stevens, "Rush to Smaller Cars Spurs Detroit to Alter Assembly Lines," *New York Times*, December 3, 1973, 48.
6. Consumers Power Company, *1974 Annual Report*, 5. Consumers Power Company, *1975 Annual Report* (Jackson, MI: Consumers Power Company, February 16, 1976), 4.
7. "U.S. GDP Growth Rate 1961–2020," Macrotrends, https://www.macrotrends.net/countries/USA/united-states/gdp-growth-rate.
8. Consumers Power Company, *1974 Annual Report*, 10.
9. Consumers Power Company, *1974 Annual Report*, 2.
10. Consumers Power Company, *1974 Annual Report*, 2.
11. Consumers Power Company, *1974 Annual Report*, 10.
12. Marc Levinson, *An Extraordinary Time* (New York: Basic Books, 2016), 74–78.
13. Consumers Power Company, *1971 Annual Report* (Jackson, MI: Consumers Power Company, February 16, 1972), 5.
14. Consumers Power Company, *1974 Annual Report*, 7.
15. "Midland N-Plant Cost Jumps 48%," *Midland Daily News*, April 12, 1975, 1.
16. "Costs Are Soaring For Power Plants," *Holland (MI) Evening Standard*, January 9, 1974, 14. Phillip L. Schneider, "N-Plant Cost Rises to Over $740 Million," *Midland Daily News*, January 9, 1974, 1.
17. Allan Sloan, "Consumers Plant Cost Up 50 Pct.," *Detroit Free Press*, April 12, 1975, 3.
18. Consumers Power Company, *1974 Annual Report*, 2.
19. Consumers Power Company, *1974 Annual Report*, 7.
20. Fred Girard, "Gas Users Pay for $73 Million Error," *Detroit Free Press*, June 21, 1974, 1–2.
21. Allan Sloan, "Customers Will Pay Dearly For Consumers' New Gas Plant," *Detroit Free Press*, February 3, 1974, 17.
22. Allan Sloan, "Consumers: A Sorry Tale of a Short-Circuited Utility," *Detroit Free Press*, June 20, 1974, 4.
23. Sloan, "Consumers," 5. At the time of this story, interstate pipelines controlled the flow of gas from the southern areas of gas production in the United States to the Midwest. Interstate pipelines purchased natural gas from producers, transported it to its customers and sold the bundled product for a regulated price. Instead of being able to purchase the natural gas as one product and the transportation as a separate service, pipeline customers were offered no option to purchase the natural gas and arrange for its transportation separately. In 1992 interstate pipelines were required to "unbundle" their service, essentially separating the sale of natural gas from the transportation. For a complete history of the regulation of natural gas, see "The Market under Regulation,"

NaturalGas.org, http://naturalgas.org/regulation/market.

24. Girard, "Gas Users Pay for $73 Million Error," 1–2.
25. Allan Sloan, "Al Aymond: The Man Under the Gun," *Detroit Free Press*, June 20, 1974, 5.
26. A good example of why the company felt they needed to continually raise their rates can be found in the 1973 book, *Future Builders*, whose author says the book was written with the "unstinting support" of Consumers Power's public relations employees and A. H. Aymond's door was always open to him. George Bush, *Future Builders: The Story of Michigan's Consumers Power Company* (New York: McGraw-Hill Book Company, 1973), x, 435–441. The general view of electric utilities in Michigan during the period of this story that any conceivable need for electricity be met is discussed in Jordon P. Howell, "Powering 'Progress': Regulation and the Development of Michigan's Electricity Landscape," *Annals of the Association American Geographers* 101, no. 4 (2011): 962–970, DOI: 10.1080/00045608.2011.569661.
27. Sloan, "Al Aymond: The Man under the Gun," 5.
28. "PSC Wants Hearings On Consumers Rates," *Lansing State Journal*, October 6, 1972, 18.
29. "Consumers' Rate Rehearing Urged," *South Bend (IN) Tribune*, February 21, 1974, 7.
30. Sloan, "Al Aymond: The Man under the Gun," 5.
31. John L. Campbell, *Collapse of an Industry* (Ithaca, NY: Cornell University Press, 1988), 3–4.
32. Paul Rau, "Are Ratepayers Protected," *Midland Daily News*, August 30, 1983, 12.
33. Consumers Power Company, *1975 Annual Report*, 8.
34. Consumers Power Company, *1975 Annual Report*, 8.
35. Allan Sloan, "The Challenge of Consumers Power Drew Selby Away from California Sun," *Detroit Free Press*, October 19, 1975, B1, B3. "New Consumers President Has Had Busy Life," *Herald-Palladium* (St. Joseph, MI), April 18, 1975, 21.
36. "Consumers Power Sues Nuclear Plant Builder," *Lansing State Journal*, August 29, 1974, 3.
37. Allan Sloan, "Consumer Gets $14 Million in A-Plant Dispute," *Detroit Free Press*, May 21, 1977, 22.
38. Richard D. James, "Bechtel: Industrial giant," *Daily Item* (Sunbury, PA), April 11, 1979, 3.
39. Allan Sloan, "Midland Power Plants Faces Test from Dow," *Detroit Free Press*, May 3, 1975, 3. "Dow Renegotiating Consumers' Pact," *News-Palladium* (Benton Harbor, MI), January 22, 1974, 9.
40. Sloan, "Midland Power Plants Faces Test from Dow," 3.
41. "Dow Uneasy about Delay," *Lansing State Journal*, November 17, 1974, 53.
42. Consumers Power Company, *1975 Annual Report*, 3.
43. Linda J. Lien, "Project Gains Momentum," *Midland Reactor* 1, no. 1, November 1976, 1.

CHAPTER 15. MIDLAND PLANT: OPPORTUNITIES (AND FRUSTRATIONS) FOR ALL

1. Betty Herring, "Nuclear Plant Called 'Greatest Investment in the State,'" *Midland Daily News*, June 27, 1969, 1, 3.
2. Starr Eby, "Nuclear Plant Right on Schedule—Utility," *Midland Daily News*, January 19, 1978, 1.
3. Linda J. Lein, "A Labor Day Salute—1978," *Midland Reactor*, September 1978, 5.
4. Consumers Power Company, *1978 Annual Report* (Jackson, MI: Consumers Power Company, February 9, 1979), 18. According to "Comments from Consumers Power," *Midland Reactor* (May 1978, 2), for the month of March 1978 Bechtel's payroll for onsite employees was $8,645,000, which was up from an average monthly payroll in 1977 of $5,851,000.
5. Paul Rau, "Dow Steam Price Was 'Intolerable,'" *Midland Daily News*, July 15, 1983, 1.
6. Tom Bradley, interview with author, April 28, 2017.
7. Bruce Peck, interview with author, April 20, 2016.
8. Linda J. Lein, "Groeschen Wins Contest for Naming Newspaper," *Midland Reactor*, November 1976, 1.
9. Linda J. Lein, "Onsite Welding Class Graduated 18 Craftsmen," *Midland Reactor*, November 1976, 3.
10. Linda J. Lein, "Focus," *Midland Reactor*, February 1977, 5. Yes, the article did refer to Uptegraff as a "lady." Although when it quoted her, she referred to herself as a woman.
11. Liz Uptegraff, telephone conversation with author, February 1, 2017, and August 18, 2018.
12. Bonnie Holley, "Focus," *Midland Reactor*, January 1981, 4.
13. Bonnie Holley, "Focus," *Midland Reactor*, April 1980, 4.
14. Chris Gallandt, telephone conversation with author, September 23, 2016.
15. Bonnie Holley, "Apprentices Learn On-the-Job," *Midland Reactor*, April 1980. 3.
16. Jan Hamann, "Focus," *Midland Reactor*, August 1982, 7.
17. Phillip L. Schneider, "N-Plant Cost Rises to Over $740 Million," *Midland Daily News*, January 9, 1974, 1.
18. "Quality Assurance at the Midland Nuclear Powerplant," Oversight Hearing Before the Subcommittee on Energy and the Environment of the Committee on Interior and Insular Affairs, House of Representatives, Ninety-Eighth Congress, First Session, on Quality Assurance at the Midland Nuclear Powerplant, Hearing Held in Washington, D.C., June 16, 1983. Testimony of John D. Selby, President and Chief Executive Officer, Consumers Power Company, 10.
19. Irving Spiewak and David F. Cope, *Overview Paper on Nuclear Power* (Oak Ridge, TN: Oak Ridge National Laboratory Report ORNL/TM-7425, September 1980), 108–109, https://

inis.iaea.org/collection/NCLCollectionStore/_Public/12/582/12582196.pdf?r=1&r=1.

20. Bernard L. Cohen, "Chapter 9: Costs of Nuclear Power Plants—What Went Wrong?," in *The Nuclear Energy Option* (New York: Plenum Press, 1990), http://www.phyast.pitt.edu/~blc/book/chapter9.html.
21. Bonnie Holly, "Comments from Consumers," *Midland Reactor*, January 1981, 2.
22. Consumers Power Company, *1977 Annual Report* (Jackson, MI: Consumers Power Company, February 10, 1978), 3.

CHAPTER 16. 1978: NOT ONLY OPTIMISM SINKS AND CRACKS

1. Consumers Power Company, *1977 Annual Report* (Jackson, MI: Consumers Power Company, February 10, 1978), 11. Consumers Power Company, *1976 Annual Report* (Jackson, MI: Consumers Power Company, February 11, 1977), 11.
2. Consumers Power Company, *1977 Annual Report*, 9.
3. Consumers Power Company, *1978 Annual Report 1978* (Jackson, MI: Consumers Power Company, February 9, 1979), 14.
4. Consumers Power Company, *1977 Annual Report*, 17.
5. "Midland N-Plant Cost Jumps 48%," *Midland Daily News*, April 12, 1975, 1. John L. Campbell, *Collapse of an Industry* (Ithaca, NY: Cornell University Press, 1988), 130–132. Since coal prices for electrical generation followed oil prices in 1973–1974—going from less than $10 per ton in November of 1973 to over $24 per ton in July 1974—the high price of uranium did not make a large change in the comparative cost of nuclear generation and coal generation of electricity. See Don Fermoyle and Kathy Goldforth, "Power Firm Unit Asks Probe of Coal Prices," *Akron (OH) Beacon Journal*, October 19, 1974, 15.
6. "Reactor Builder's Goof," *Lansing State Journal*, April 21, 1977, 22. Tendon sheaths are tubes through which tension cables are strung to ensure the strength of the containment structure.
7. David Everett, "Decade of Mistakes at Midland A-Plant," *Detroit Free Press*, November 30, 1981, 13A.
8. Consumers Power Company, *1977 Annual Report*, 1.
9. Consumers Power Company, *1979 Annual Report* (Jackson, MI: Consumers Power Company, February 15, 1980), 13. Consumers Power Company, *1984 Annual Report* (Jackson, MI: Consumers Power Company, February 1, 1985), 3.
10. "Anderson: Utility Should Pay for Errors," *Midland Daily News*, August 30, 1983, 11.
11. Bonnie Holley, "84 Ton Lift—Diesel Generators Move In," *Midland Reactor*, September 1980, 1.

12. Starr Eby, "NRC, Utility to Inspect Nuclear Plant Settling," *Midland Daily News*, December 2, 1978, 1.
13. Allan Sloan, "Building for Backup Unit at Nuclear Plant Is Sinking," *Detroit Free Press*, November 29, 1978, 44.
14. Shun-ichi Tanaka, "Accident at the Fukushima Dai-ichi Nuclear Power Stations of TEPCO—Outline & Lessons Learned—," *Proceedings of the Japan Academy, Series B: Physical and Biological Sciences* 88, no. 9 (2012): 471–484. doi:10.2183/pjab.88.471. P. Y. Lipscy, K. E. Kushida, and Trevor Icerti, "The Fukushima Disaster and Japan's Nuclear Plant Vulnerability in Comparative Perspective," *Environmental Science & Technology* 47 (2013): 6082, doi: https://doi.org/10.1021/es4004813.
15. "Fukushima Daiichi Accident," World Nuclear Association, updated October 2018, https://www.world-nuclear.org/information-library/safety-and-security/safety-of-plants/fukushima-accident.aspx.
16. *Interim Report: Investigation Committee on the Accident at Fukushima Nuclear Power Stations of Tokyo Electric Power Company; Investigation Committee on the Accident at Fukushima Nuclear Power Stations of Tokyo Electric Power Company* (Tokyo, December 26, 2011), 2, https://www.cas.go.jp/jp/seisaku/icanps/eng/interim-report.html. By a Cabinet Decision on May 24, 2011, the Japanese government, established the Investigation Committee on the Accident at Fukushima Nuclear Power Stations of Tokyo Electric Power Company. The aim was to make policy proposals on measures to prevent further spread of the damage caused by the accident and to prevent a recurrence of similar accidents in the future by conducting a multifaceted investigation in an open and neutral manner that is accountable to the Japanese public to determine the causes of the accident at the Fukushima Dai-ichi and Dai-ni Nuclear Power Stations.
17. Tanaka, "Accident at the Fukushima Dai-ichi," 471.
18. Tanaka, "Accident at the Fukushima Dai-ichi," 473.
19. Mari Yamaguchi, "Japan Revises Fukushima Cleanup Plan, Delays Key Steps," *U.S. News & World Report*, December 27, 2019, https://www.usnews.com/news/business/articles/2019-12-26/japan-revises-fukushima-cleanup-plan-delays-key-steps. Will Croxton and Brit McCandless Farmer, "Fukushima's Ghost Towns," *60 Minutes Overtime*, 2019 CBS Interactive Inc., November 25, 2018, https://www.cbsnews.com/.
20. Starr Eby, "NRC, Utility to Inspect Nuclear Plant Settling," *Midland Daily News*, December 2, 1978, 1.
21. U.S. Nuclear Regulatory Commission, *IE Circular No. 81-08, Foundation Materials* (Washington, D.C.: U.S. Nuclear Regulatory Commission, 1981), 1. https://www.nrc.gov.
22. Rodney Wieland, discussion with author, May 26, 2016. The Tittabawassee River was not

a candidate for a cooling source because this would have added the discharge of heat into the river (thermal pollution), and during part of the year, the flows were too low to be used for cooling. Using only cooling towers was ruled out because of the fog this would have been created.

23. Consumers Power Company, *Environmental Report Operating License Stage, Revision* 1, April 1978, Volume 1, 2.4–5. This document is available from the Midland Historical Society.
24. Consumers Power Company, *Environmental Report Operating License Stage*, 2.4–5.
25. Consumers Power Company, *Environmental Report Operating License Stage*, 2.4–5.
26. "Tittabawassee River at Midland," Advanced Hydrologic Prediction Service, National Weather Service, https://water.weather.gov/ahps2/river.php?wfo=dtx&wfoid=18774&riverid=203834&pt%5B%5D=143667&allpoints=143667&data%5B%5D=crests. In an email to the author on December 27, 2020, Thomas L. Weaver, Chief, Hydrologic Date-East of the U.S, Geologic Survey said that the recorded peak elevation of the Tittabawassee on May 20, 2020 was 614.20 feet. He noted that in May of 2020 the elevation of Lake Huron was 582 feet.
27. Consumers Power Company, *Environmental Report Operating License Stage*, 2.5–3.
28. Linda Lein, "Project Gains Momentum," *Midland Reactor*, November 1976, 1, 4.
29. Lorie Shane, "N-Plant Reinforcement Plan Presented," *Midland Daily News*, December 2, 1981, 3.
30. U.S. NRC, *IE Circular No. 81-08*, 1.
31. Paul Rau, "Soil Defect Cause Probed," *Midland Daily News*, July 9, 1981, 3.
32. "Excerpts from Suit," *Midland Daily News*, July 15, 1983. This article reports the twenty-nine–page lawsuit filed on July 14, 1983 in Midland Circuit Court by the Dow Chemical against Consumers Power.
33. Eby, "NRC, Utility to Inspect Nuclear Plant Settling," 1.
34. Paul Rau, "Underpinning N-Plant Is a Slow, Complex Task," *Midland Daily News*, September 5, 1983, 9.
35. Paul Rau and Lorie Shane, "Soils Work Can Begin, NRC Says," *Midland Daily News*, December 10, 1982, 1.
36. Bob Wheeler, discussion with author, May 25, 2016.
37. J. Samuel Walker and Thomas R. Wellock, *A Short History of Nuclear Regulation 1946–2009* (Washington, D.C.: Office of the Secretary U.S. Nuclear Regulatory Commission, NUREG/BR-0175, Rev. 2, October 2010), 52, https://www.nrc.gov/docs/ML1029/ML102980443.pdf.
38. Norman C. Rasmussen, *Reactor Safety Study: An Assessment of Accident Risks in*

U. S. Commercial Nuclear Power Plants (Washington, D.C.: U.S. Nuclear Regulatory Commission, WASH-1400-MR; NUREG-75/014-MR, 1975), https://www.osti.gov/biblio/7134131.

39. Walker and Wellock, *A Short History of Nuclear Regulation 1946–2009*, 53.
40. Allan Sloan, "The Challenge of Consumers Power Drew Selby Away from California Sun," *Detroit Free Press*, October 19, 1975, 19.
41. "Selby Wears 3 Hats for Consumers Power," *Lansing State Journal*, October 4, 1979, B6.

CHAPTER 17. TMI2: DEFINITELY NOT TOO MUCH INFORMATION

1. Bud Farrar, "Dow Project Challenged in Company Town," *Edwardsville (IL) Intelligencer*, April 20, 1974, 2. It seems clear that Dr. North believed that the chances of a major nuclear accident were one in a billion and not a billion to one as reported.
2. *The China Syndrome*, directed by James Bridges (Chatsworth, CA: Columbia Pictures, 1979).
3. J. Samuel Walker, *Three Mile Island: A Nuclear Crisis in Historical Prospective* (Berkeley: University of California Press, 2004), 2.
4. In 1986 an example of the real possibility of a large geographic area being rendered radioactive with the meltdown of a nuclear reactor did occur with the Chernobyl accident. Adam Higginbotham writes that the melted core of the Chernobyl reactor sat just a few meters above the water table of the Pripyat River. If the core had penetrated to the water table, the drinking water of Kiev would have been poisoned as well as that of everyone in Ukraine who drew waters of the Dnieper River basin—some thirty million people. A second danger at Chernobyl was that, if the molten fuel had gotten into flooded safety compartments beneath the reactor floor, there could have been a steam explosion that would have hurled fallout into the atmosphere and rendered a large swath of Europe uninhabitable for a hundred years. The Chernobyl reactor was a fundamentally different design from those at Three Mile Island, Midland, and most of the reactors in the world, but it has become the modern poster child for opposition to nuclear power. Adam Higginbotham, *Midnight in Chernobyl* (New York: Simon & Schuster, 2019), 191–192.
5. Walker, *Three Mile Island*, 78–82.
6. John G. Kemeny, "Account of the Accident: March 28, 1979," Report of the President's Commission on the Accident at Three Mile Island, http://www.pddoc.com/tmi2/kemeny/wednesday_march_28_1979.htm. This account contains the commission staff notes of Pintek's interview.
7. William Lanouette, "Three Mile Island Plus 10: Will Press Coverage Be Better Next Time,"

C-Span, March 28, 1989, https://www.c-span.org/video/?6883-1/mile-island-press.

8. Dan Nimmo, "TV Network News Coverage of Three Mile Island: Reporting Disasters as Technological Fables," *International Journal of Mass Emergencies and Disasters* 2, no. 1 (March 1984): 128.
9. J. Samuel Walker and Thomas R. Wellock, *A Short History of Nuclear Regulation, 19462009* (Washington, D.C.: U.S. Nuclear Regulatory Commission, NUREG/BR-0175, Rev. 2, October 2010), 140. http://www.nrc.gov/docs/ML1029/ML102980443.pdf.
10. Bob Dvorchak, "Steam Escapes at N Plant," *Midland Daily News*, March 28, 1979, 1.
11. Bob Dvorchak, "Radiation Leak Cause Is Debated," *Midland Daily News*, March 29, 1979, 1.
12. "Accident Over in 8 Minutes," *Midland Daily News*, March 29, 1979, 20.
13. Starr Eby, "Will Leak Effect Midland Plant?," *Midland Daily News*, March 30, 1979, 1. Sinclair's reference on March 30th to the public notice occurring at 7 a.m. apparently refers to 7:05 a.m. on March 28th when Gary Miller, the Three Mile Island station manager, declared a general emergency, defined as having the "potential for serious radiological consequences for the health and safety of the general public." See Kemeny, "Account of the Accident: March 28, 1979," fn. 49–51, which puts the notice to the public as happening at 8:25 a.m.
14. Bob Dvorchak, "Persons Near Plant Told to Stay Inside, *Midland Daily News*, March 30, 1979, 1.
15. Bob Dvorchak, "N-Plant Bubble Reduced," *Midland Daily News*, March 31, 1979, 1.
16. Paul Rau, "More Safeguards for Plant Here," *Midland Daily News*, April 2, 1979, 1.
17. Bob Dvorchak, "Bubble Shrinking; Temperature Drops," *Midland Daily News*, April 2, 1979, 1.
18. Bob Dvorchak, "Reactor Future Doubtful," *Midland Daily News*, April 3, 1979, 1.
19. Bob Dvorchak, "Catastrophe Threat Over," *Midland Daily News*, April 4, 1979, 1.
20. John A. Palen, "Nuclear Questions Plentiful, Answers Few," editorial, *Midland Daily News*, April 4, 1979, 4.
21. Kemeny, "Account of the Accident."
22. Walker, *Three Mile Island*, 209–215.
23. Starr Eby, "Startup Delay Possible," *Midland Daily News*, April 3, 1979, 1.
24. Eby, "Startup Delay Possible," 1.
25. Nimmo, "TV Network News Coverage of Three Mile Island: Reporting Disasters as Technological Fables," 128.

CHAPTER 18. PEOPLE POWER: FOUR THOUSAND MARCH

1. Constance E. Cook, *Nuclear Power and Legal Advocacy* (Lexington, MA: Lexington Books, 1980), 11.
2. Ralph Nader and John Abbotts, *The Menace of Atomic Energy* (New York: W.W. Norton, 1977), 59, 107.
3. Cook, *Nuclear Power and Legal Advocacy*, 51–52.
4. Sara Hughes, telephone interview with author, September 1, 2016.
5. Gerard H. Clarfield and William M. Wiecek, *Nuclear America: Military and Civilian Nuclear Power in the United States, 1940–1980* (New York: Harper & Row, 1984), 218.
6. Timothy J. Jorgensen, *Strange Glow* (Princeton, NJ: Princeton University Press, 2016), 189. See also Robert P. Crease, "Fallout," in *Science and Other Cultures: Issues in Philosophies of Science and Technology*, ed. Robert Figueroa and Sandra Harding (New York: Routledge, 2003), 106.
7. Terrence R. Fehner and F. G. Gosling, *Battlefield of the Cold War: The Nevada Test Site; Atmospheric Nuclear Weapons Testing 1951–1963* (Washington, D.C.: U.S. Department of Energy, 2006), 114–116, https://www.energy.gov/sites/prod/files/DOENTSAtmospheric.pdf.
8. Hughes, telephone interview.
9. Hughes, telephone interview.
10. Terry R. Miller, "Environmental Justice: A History of the Lone Tree Council's Grassroots Effort," *State History Conference*, Saginaw Valley State University, September 26, 2015, 1.
11. Patrick Race and Terry Miller, interview with author, January 17, 2017.
12. Kathy Roberts, "The Lone Tree," *Bay City (MI) Times*, June 30, 1998, 1D.
13. Leslie E. Arndt, *The Bay County Story* (Detroit: Harlo Printing Company, 1982), 32.
14. Race and Miller, interview.
15. Jerry Morlock, "Nuke Protesters in Midland," *Central Michigan Life* (Mount Pleasant, MI), November 20, 1978, 12.
16. Morlock, "Nuke Protestors in Midland," 12.
17. "Michael Moore, Full Biograph," *New York Times*, February 10, 2010.
18. "Flint Voice," *Flint (MI) Voice* 3, no. 3, April 1979, 3.
19. "Stop Nuclear Power!," *Flint (MI) Voice* 3, no. 3, April 1979, 1.
20. Starr Eby, "3,000 May March on Saturday," *Midland Daily News*, April 19, 1979, 3.
21. "Expect Record Number to Protest in Midland," *Ludington (MI) Daily News*, April 19, 1979, 2.
22. "Expect Record Number to Protest in Midland," 2.
23. John A. Palen, "Rain Fails to Dim Marchers' Spirit," *Midland Daily News*, April 23, 1979, 3.

24. "Whiting Overlook Park," Midland Area Community Foundation, https://www.midlandfoundation.org/whiting-overlook-park/.
25. Doug Hall, "4,000 Nuclear Power Foes March on Midland Reactor," *Detroit Free Press*, April 22, 1979, 3.
26. Eby, "3,000 May March on Saturday," 1. The Silkwood case was widely used to illustrate the safety and management problems of nuclear power. Karen Silkwood worked at a Kerr-McGee plant in Oklahoma that fabricated plutonium fuel for use in nuclear power plants. She raised concerns about the safety of the plant, and on her way to meet a *New York Times* reporter, she was killed in a one-car accident on November 13, 1974. It was later alleged that she was forced off the road in the incident that led to her death. Her father brought a negligent conduct suit against Kerr-McGee for failure to train their employees and failure in their handling of plutonium. The case was eventually settled out of court for $1.38 million with Kerr-McGee admitting no liability. In the 1983 movie *Silkwood*, Meryl Streep played the role of Silkwood. At the 56th Academy Awards, *Silkwood* received five nominations in total, including Streep for Best Actress and Cher for Best Supporting Actress. For more information, see George J. Annas, "The Case of Karen Silkwood," *Public Health and Law* 74, no. 5 (May 1984): 516–518, https://www.ncbi.nlm.nih.gov/pmc/articles/PMC1651625/pdf/amjph00628-0100.pdf.
27. "Anti-nuclear Protest Peaceful in Midland; Leak Hits Big Rock Plant," *Battle Creek (MI) Enquirer*, April 22, 1979, 2.
28. Palen, "Rain Fails to Dim Marchers' Spirit," 3.
29. Hall, "4,000 Nuclear Power Foes March on Midland Reactor," 3.
30. "4,000 Protest at Nuclear Plant," *Chicago Tribune*, April 22, 1979, 59.
31. Michael Hayes, interview with author, November 27, 2017.
32. Donna Sanks, "Nuclear Plant Critic Ponders Election Bid," *Midland Daily News*, April 22, 1980, 1.
33. Fred E. Garrret, "'Hot Race' in 102nd Ends Up as a Walk-Away," *Saginaw News*, November 5, 1980, 1.
34. Hayes, interview.

CHAPTER 19. THE 1980S: PROBLEMS INTENSIFY AS THE MONEY DRAINS

1. "Dow Drops Its Support of Midland Nuclear Plant," *Argus-Press* (Owosso, MI), February 3, 1977, 11.
2. "Dow Drops Its Support of Midland Nuclear Plant," 11.
3. "Dow Says Jobs Hang on Nuclear Power," *Lansing State Journal*, February 4, 1977, 10.

4. Allan Sloan, "Pact over Midland A-Plants appears Tilted Towards Dow," *Detroit Free Press*, June 28, 1978, 47.
5. Paul Rau, "N-Plant Soil Problems to Receive Hearing," *Midland Daily News*, March 26, 1980, 3.
6. Lorrie Shane and Paul Rau, "Panel Accepts Part of NRC, Consumes Pact," *Midland Daily News*, July 8, 1981, 3.
7. Paul Rau, "Mrs. Stamiris Ending Major Role," *Midland Daily News*, December 26, 1981, 3.
8. Paul Rau, "NRC Sees Stipulations as Helpful to Hearings," *Midland Daily News*, July 23, 1981, 3.
9. Shane and Rau, "Panel Accepts Part of NRC, Consumes Pact," 3.
10. Paul Ray, "Cavern at N-Plant Will Be Monitored," *Midland Daily News*, July 11, 1981, 1. The location of the wells Dow Chemical used to extract salt at depths between 4,100 and 4,300 feet below the surface are shown in figure 2.5–5 in *Midland Plant—Units 1 and 2, Environmental Report, Operating License Stage*, vol. 1, April 1978 (Midland: Midland County Historical Society).
11. Rodney Wieland, interview with author, May 26, 2016.
12. "Obituary for James Arthur Mooney," *Blount Countian* (Oneonta, AL), June 10, 2009.
13. Rodney Wieland, discussion with author, May 31, 2016. Wieland was an engineer who worked for Consumers Power on the Midland Plant.
14. "The Staff of the Public Service Commission, Concerned," UPI Archives, April 16, 1984, https://www.upi.com/Archives/1984/04/16/The-staff-of-the-Public-Service-Commission-concerned-over/4590450939600/.
15. David Everett, "Consumers Digs to Stop Sinking at Nuclear Plant," *Detroit Free Press*, April 15, 1984, 1A, 14A.
16. Paul Rau, "Underpinning N-Plant Is a Slow, Complex Task," *Midland Daily News*, September 5, 1983, 9.
17. David Everett, "What Began As a Dream . . . ," *Detroit Free Press*, July 18, 1984, 15.
18. Charlotte W. Craig, "Midland Nuke Plants, Now $3.1 Billion, Still On," *Detroit Free Press*, March 7, 1980, 36.
19. John R. Emshwiller and Robert L. Simison, "Midwest Nuclear Plant, in the Works 13 Years, Keeps Facing Delays," *Wall Street Journal*, March 4, 1980, 1.
20. Consumers Power Company, *1980 Annual Report* (Jackson, MI: Consumers Power Company, February 13, 1981), 13.
21. Consumers Power Company, *1983 Annual Report* (Jackson, MI: Consumers Power Company, February 3, 1984), 21.
22. Year-end total expense for the Midland Plant can be found on the following pages of the

annual reports: 1980 on page 13, 1981 on page 12, 1982 on page 17, 1983 on page 21, and 1984 on page 11. The yearly expense totals reported are the differences in total expense from one year to the next. Consumers Power Company, *1980 Annual Report*; Consumers Power Company, *1981 Annual Report* (Jackson, MI: Consumers Power Company, February 12, 1982); Consumers Power Company, *1982 Annual Report* (Jackson, MI: Consumers Power Company, February 4, 1983); Consumers Power Company, *1983 Annual Report*; Consumers Power Company, *1984 Annual Report* (Jackson, MI: Consumers Power Company, February 1, 1985).

23. "Consumer Price Index, 1913–," Federal Reserve Bank of Minneapolis, https://www.minneapolisfed.org/about-us/monetary-policy/inflation-calculator/consumer-price-index-1913-.
24. "GDP Growth (Annual %)—United States," The World Bank, https://data.worldbank.org/indicator/NY.GDP.MKTP.KD.ZG?locations=US.
25. "Michigan Again Sets the Pace in Unemployment." *Lawrence (KS) Journal Word*, November 16, 1982, 22.
26. "State's Jobless Rate Trails Only Michigan," *Gadsden (AL) Times*, January 19, 1983, 18.
27. Bruce Peck, interview with author, April 20, 2016.
28. Roderick S. (Rick) Coy, telephone conversation with author, November 11, 2017. Coy started as an intern in Frank Kelley's office in 1970 and was Assistant Attorney General of Michigan from 1971 to 1981 where he concentrated in electric, gas, and communications utility regulation and administrative litigation.
29. Consumers Power Company, *1980 Annual Report*, 7. Consumers Power Company, *1967 Annual Report* (Jackson, MI: Consumers Power Company, February 16, 1968), 19.
30. Consumers Power Company, *1982 Annual Report*, 17.
31. Consumers Power Company, *1981 Annual Report*, 13.
32. Consumers Power Company, *1981 Annual Report*, 13.
33. Consumers Power Company, *1981 Annual Report*, 13.
34. Consumers Power Company, *1983 Annual Report*, 2.
35. "Everything You Need to Know about Junk Bonds," Investopedia, August 17, 2019, https://www.investopedia.com/articles/02/052202.asp.
36. Jan Hamann, "Oakway IV Funds Midland Construction," *Midland Reactor*, February 1982, 8.
37. Consumers Power Company, *1982 Annual Report*, 18.
38. Consumers Power Company, *1982 Annual Report*, 16.
39. "Boris Sure He Can Raise Money Needed to Finish Nuclear Plant," *Midland Daily News*, August 13, 1983, 13.

40. Consumers Power Company, *1983 Annual Report*, 15, 17.
41. Consumers Power Company, *1983 Annual Report*, 17.
42. Pat McCarthy, "Nuclear Power Best Fuel Available, Says Consumers Head," *Battle Creek (MI) Enquirer*, March 15, 1980, A-4.
43. Consumers Power Company, *1981 Annual Report*, 4.
44. Consumers Power Company, *1981 Annual Report*, 2.
45. Consumers Power Company, *1981 Annual Report*, 2.
46. Consumers Power Company, *1981 Annual Report*, 2.
47. Consumers Power Company, *1983 Annual Report*, 3.
48. "N-Plant Bill May Go Up by $1.5 Billion," *Midland Daily News*, April 10, 1984, 1.

CHAPTER 20. QUALITY CONTROL PROBLEMS: DÉJÀ VU

1. David Everett, "Consumers Power Bets All on Nuke Plant," *Detroit Free Press*, June 20, 1983, 1A.
2. Lorie Shane, "Foundation Adequate, Engineer Testifies," *Midland Daily News*, December 8, 1982, 3. Everett, "Consumers Power Bets All on Nuke Plant," 15A.
3. "Nuclear Power Plants / Midland, Michigan," *CBS Evening News*, Dan Rather and Robert Schakne, December 14, 1982, Vanderbilt TV News Archive, https://tvnews.vanderbilt.edu/broadcasts/281614. At this time, the *CBS Evening News* was the most watched early evening news program in the United States. Julianne Hastings, "TV World; NEWLN: CBS Top Network for 1983–84 in Four Ratings Categories," UPI Archives, April 18, 1984, https://www.upi.com/Archives/1984/04/18/TV-WorldNEWLNCBS-top-network-for-198384-in-four-ratings-Categories/7224451112400/. Richard K. Shull, "Success in Small Numbers," *Indianapolis News*, November 17, 1982, 23.
4. "Nuclear Power Plants / Midland, Michigan."
5. Terry R. Miller, "Environmental Justice: A History of the Lone Tree Council's Grassroots Effort" paper presented at the State History Conference, SVSU, Saginaw Valley State University, University Center, MI, September 26, 2015, 2.
6. "Quality Assurance at the Midland Nuclear Powerplant," Oversight Hearing Before the Subcommittee on Energy and the Environment of the Committee on Interior and Insular Affairs, House of Representatives, Ninety-Eighth Congress, First Session, on Quality Assurance at the Midland Nuclear Powerplant, Hearing Held in Washington, D.C., June 16, 1983, 126 (statement of Billie Pirner Garde, director, Citizens Clinic, Government Accountability Project, Institute of Policy Studies).
7. Dennis Pfaff, "Midland Construction Blasted as Unsafe," UPI Archives, June 29,

1982, https://www.upi.com/Archives/1982/06/29/Midland-construction-blasted-asunsafe/2802394171200/.

8. "Quality Assurance at the Midland Nuclear Powerplant," 121–136. During my research I was given copies of two of the affidavits taken by GAP's David Crow and Billie Garde that were part of this investigation of the Midland Plant. These documents came from the estate of Sarah Hughes (1933–2018). Prior to her death, I talked with her on several occasions, and I believe these documents to be copies of the original affidavits. I can't vouch for the veracity of the claims made in the affidavits, but the information in them is consistent with the electrician's account on *CBS Evening News* and with the comments of other individuals I interviewed. Hughes worked for the law firm of Sinclair and Clio and also assisted Mary Sinclair with typing and office work.
9. The name of the person who provided the affidavit was cut out of the copy provided to me.
10. Dennis M. King, letter to the editor, "Engineer Questions Article On Midland Nuclear Plant," *Herald-Palladium* (St. Joseph, MI), July 10, 1982, 7.
11. Julie Morrison, "9 Violations at N-Plant Called Minor," *Midland Daily News*, March 28, 1984, 3.
12. "9 Arrested on Drug Charges at Midland Nuclear Plant," *Ludington (MI) Daily News*, September 8, 1983, 3.
13. Bruce Peck, interview with author, August 20, 2016.
14. "9 Arrested on Drug Charges at Midland Nuclear Plant," 3. Kathy Gray, "Arraignments Set for Nine," *Midland Daily News*, October 8, 1983, 3.
15. "9 Arrested on Drug Charges at Midland Nuclear Plant," 3.
16. "9 Held in Midland A-plant Drug Raid," *Detroit Free Press*, September 8, 1983, 9.
17. Jan Hamann, "Midland QIP Set in Motion," *Midland Reactor*, October 1981, 1,3.
18. Jan Hamann, "Quality Corner," *Midland Reactor*, April 1982, 5.
19. Chris Aldridge, "Flooding Damages Midland Cogeneration Venture," *Midland Daily News*, June 30, 2017, https://www.ourmidland.com/news/article/Flooding-damages-Midland-Cogeneration-11257847.php#item-85307-tbla-2.
20. "Tittabawassee River at Midland," Advanced Hydrologic Prediction Service, National Weather Service, https://water.weather.gov/ahps2/river.php?wfo=dtx&wfoid=18774&riverid=203834&pt%5B%5D=143667&allpoints=143667&data%5B%5D=crests.
21. James Rajewski, telephone interview with author, March 4, 2017. Rajewski is the former chief financial officer of the Midland Cogeneration Venture Limited Partnership.
22. Lorie Shane, "Foundation Adequate, Engineer Testifies," *Midland Daily News*, December 8, 1982, 3.

23. Jan Hamann, "Project Implements New CCP," *Midland Reactor*, December 1982, 2.
24. "Quality Assurance at the Midland Nuclear Powerplant," 1.
25. "Quality Assurance at the Midland Nuclear Powerplant," 17–18.
26. "Quality Assurance at the Midland Nuclear Powerplant," 18.
27. "Quality Assurance at the Midland Nuclear Powerplant," 35.
28. "Quality Assurance at the Midland Nuclear Powerplant," 35.
29. "Quality Assurance at the Midland Nuclear Powerplant," 35.
30. "Quality Assurance at the Midland Nuclear Powerplant," 35.
31. David Everett, "Regulators Back Midland Project," *Detroit Free Press*, April 26, 1984, 1A, 12A.
32. "Consumers Finds Ally in NRC," *Lansing State Journal*, April 26, 1984, 4B.
33. Everett, "Regulators Back Midland Project," 11.

CHAPTER 21. BRINK OF BANKRUPTCY: NO SALES CONTRACT, NO COST RECOVERY, AND NO MONEY

1. Jeff Mayers, "Midland Nuclear Plant: A Generator of Problems," *Midland Daily News*, October 14, 1983, 3. Walter Boris's official title at Consumers Power was executive vice president of financial and corporate affairs.
2. Mayers, "Midland Nuclear Plant," 3.
3. Paul Rau, "Dow: Steam Price Was 'Intolerable,'" *Midland Daily News*, July 15, 1983, 1.
4. "State to Fight Consumers' Rate Increase," *Lansing State Journal*, November 19, 1983, 1. This case was U-7830. The total increase requested for the Midland Plant was $564 million, but Consumers Power said fuel savings from the Midland Plant would be $73 million per year or more. The American Association of Businesses Advocating Tariff Equity (ABATE) was an industrial organization that included the Big Three automakers (General Motors, Ford, and Chrysler) and other large Michigan businesses. For more about the opposition to Consumers Power plans to raise its rates, see David Everett, "What Began as a Dream," *Detroit Free Press*, July 18, 1984, 15.
5. Allan Sloan, "Consumers: A Sorry Tale of a Short-Circuited Utility," *Detroit Free Press*, June 20, 1974, 5.
6. "Mistakes Were Honest," *Midland Daily News*, August 31, 1983, 13.
7. Frank J. Kelley and Jack Lessenberry, *The People's Lawyer* (Detroit: Wayne State University Press, 2015), 66.
8. Kelley and Lessenberry, *The People's Lawyer*, 5, 73–75, 84, 110–111, 172–175.
9. Rodney Boulanger, interview with author, December 14, 2016.

10. Kelley and Lessenberry, *The People's Lawyer*, 83.
11. "Anderson: Utility Should Pay for Errors," *Midland Daily News*, August 30, 1983, 11. According to sales of electricity reporting in Consumers Power's annual reports, their sales were 24.1 billion kilowatt hours for 1973 and 24.6 billion kilowatt hours for 1982 (sales for 1982 were reported in the 1984 annual report on page 3). Based on Consumers Power's annual reports, the total sales growth from 1973 to 1982 was 2 percent. Consumers Power's sales rose in 1983 to 25.8 billion kilowatt hours and in 1984 to 26.9 billion kilowatt hours (reported on page 3 of the 1984 annual report). Consumers Power Company, *1973 Annual Report* (Jackson, MI: Consumers Power Company, February 8, 1974), 1. Consumers Power Company, *1984 Annual Report* (Jackson, MI: Consumers Power Company, February 1, 1985), 3.
12. "Anderson: Utility Should Pay for Errors," 11. I could find no data to document that Consumers Power's industrial rates were 30 percent above the Midwest average, but data published for 1983 by the U.S. Department of Energy (DOE) / U.S. Energy Information Administration (EIA) does show that industrial rates in Cincinnati and Columbus, Ohio (5.2 and 4.79 cents/KWh), were substantially below the 6.2 cents/KWh charged by Detroit Edison in Detroit. DOE/EIA-0348(85), *Electric Power Annual 1983* (Washington, D.C.: U.S. Energy Information Administration, 1984), 309.
13. John C. Carlson, letter to the editor, *Midland Daily News*, February 18, 1985, 4.
14. Paul Rau, "Anderson: Utility Should Pay for Error," *Midland Daily News*, August 30, 1983, 11.
15. Paul Rau, "Are Ratepayers Protected?," *Midland Daily News*, August 30, 1983, 12.
16. David Everett, "What Began As a Dream Died As A . . . ," *Detroit Free Press*, July 18, 1984, 15.
17. Paul Rau and Lorie Shane, "Soils Work Can Begin, NRC Says," *Midland Daily News*, December 10, 1982, 1.
18. David Everett, "Consumers Digs to Stop Sinking at Nuclear Plant," *Detroit Free Press*, April 14, 1984, 1.
19. "N-Plant Bill May Go up by $1.5 Billion," *Midland Daily News*, April 10, 1984, 1. "Midland: Abandonment Is the Best Option for the Troubled Plant," *Detroit Free Press*, April 2, 1984, 16A.
20. "Deadline Nears for Midland Talks," *Battle Creek (MI) Enquirer*, June 12, 1984, 2.
21. Mark Berns, "Loans Depend on Midland Decision," *South Bend (IN) Tribune*, May 2, 1984, 2.
22. Bernie Shellum, "Banks Balk on Consumers Loan," *Detroit Free Press*, May 3, 1984, 26.
23. Berns, "Loans Depend on Midland Decision," 2.
24. Berns, "Loans Depend on Midland Decision," 2.

25. "Deadline Nears for Midland Talks," 2.
26. Kathy Gray, "Midland Produces First Electricity," *Midland Daily News*, June 14, 1984, 1.
27. Julie Morrison, "N-Plant Working Overtime to Roll Turbine," *Midland Daily News*, June 11, 1984, 1.
28. Tim Bovee, "Utility Hoping to Find a Way to Finish Plant," *Daily Press* (Newport News, VA), June 16, 1984, 36.
29. Tim Jones, "Sen. Engler Earns Political Dividends," *Detroit Free Press*, June 27, 1984, 9.

CHAPTER 22. 1985: WHAT HAPPENED NEXT?

1. Bernie Shellum and David Everett, "Consumer Cancels $5.6 Billion Project," *Detroit Free Press*, July 17, 1984, 1A.
2. I worked with Jim Mooney from 1988 to 2004 at the Midland Cogeneration Venture Limited Partnership (MCV). Mooney was the vice president of engineering and I held the titles of vice president of gas supply and later vice president of energy supply and marketing. This statement is from my recollections from what Mooney said on numerous occasions.
3. C. E. Gatlin Jr., Gerald C. Vellender, and James A. Mooney, "Repowering of the Midland Nuclear Station," paper presented at the Tenth Annual Industrial Energy Technology Conference, Houston, TX, September 13–15, 1988, http://hdl.handle.net/1969.1/92393.
4. "Relief Laments Over End of Plant," *Detroit Free Press*, July 18, 1984, 14A.
5. John Holusha, "Decision to Halt Nuclear Project in Michigan Brings Hardships," *New York Times*, July 19, 1984.
6. Dorothy Langdon Yates, *Salt of the Earth* (Midland: The Midland Historical Society and the Center for the Arts, 1987), 233.
7. Mike Gronski, discussion with author, June 10, 2016. As I listened to Gronski, I could only think that probably he represented the point of view of many who thought that the Lone Tree Council people shot the spirit owl that looked after the people in the Saginaw Valley, and that history should start when H. H. Dow arrived in Midland in 1890 with the spirit of progress.
8. Frank J. Kelley, letter to the editor, *Battle Creek (MI) Enquirer*, July 28, 1984, 6.
9. Todd S. Purdum, "Business People: Chief Executive Resigns at Consumers Power," *New York Times*, March 7, 1985.
10. Consumers Power Company, *1984 Annual Report* (Jackson, MI: Consumers Power Company, February 1, 1985), 21.
11. Doris A. Fuller, "Bechtel Power Adjusts to Drop in Workload," *Los Angeles Times*, March 4,

1985, 43, 46.

12. Sally Denton, *The Profiteers, Bechtel and the Men Who Built the World* (New York: Simon & Schuster, 2016), 202–203, 206–208, 306. Matthew Rees, "Building the American Empire," *Wall Street Journal*, March 9, 2016, A13, https://www.wsj.com/articles/building-the-american-empire-1457478240.
13. Purdum, "Business People."
14. Julie Morrison, "Walter Boris to Retire 6 Months Early from CPCo," *Midland Daily News*, March 27, 1985, 1.
15. Jeri Kornegay, "John D. Selby: Reflections," *Jackson (MI) Citizen Patriot*, September 29, 1985, 1.
16. "Director Quits Board," *Lansing State Journal*, November 7, 1985, 19.
17. Gina Shaw, *Building for the Future: The Story of Consumers Energy 1886–2016* (St. Louis, MO: Essex Publishing Group, 2017), 60.
18. Consumers Power Company, "Letter to Shareholders," in *1984 Annual Report*, 1.
19. Bernie Shellum, "Consumers Creditors Take New Moves to Protect Their Interests," *Detroit Free Press*, December 9, 1984, 1E, 11E.
20. Tim Kiska, "Blanchard Appoints a Blue-Ribbon Panel to Study Fiscal Crisis," *Detroit Free Press*, December 10, 1982, 3. Fred Marc Biddle, "Utility Hopes to Step on Gas, Save the Day," *Chicago Tribune*, May 26, 1986.
21. Biddle, "Utility Hopes to Step on Gas, Save the Day."
22. "William T. McCormick, Jr.: CMS Energy," National Museum of American History, Smithsonian Institution, https://americanhistory.si.edu/powering/ohi/mccormk/mccorml.htm.
23. Michael Arndt, "Sale of Nuclear Plant Links Utility, Builder," *Chicago Tribune*, October 8, 1987. The cost of construction the Palisades Plant is from Gina Shaw, *Building for the Future: The Story of Consumers Energy, 1886–2016* (St. Louis, MO: Essex Publishing Group, Inc., 2017), 54.
24. "Consumers Selling Palisades Plant," *Herald-Palladium* (St. Joseph, MI), October 8, 1987, 1.
25. Arndt, "Sale of Nuclear Plant Links Utility, Builder." "Consumers Selling Palisades Plant."
26. Arndt, "Sale of Nuclear Plant Links Utility, Builder."
27. Thomas W. Lippman, "Bechtel, Westinghouse Buy into Nuclear Plant," *Washington Post*, July 20, 1990.
28. "Dow, Consumers Suits Misstated," *Lansing State Journal*, October 11, 1984, 13.
29. Marilyn Findlay, "Business Leaders Say N-Plant Loss Will Hurt," *Midland Daily News*, July 17, 1984, 10.

CHAPTER 23. CONSUMERS POWER TO CONSUMERS ENERGY: FROM NUCLEAR TO NATURAL GAS AND BEYOND

1. Tim Bovee, "Midland Nuclear Plant Dispute Deserves a Soap Opera Slot," *South Bend (IN) Tribune*, December 16, 1985, 10.
2. George White, "Consumers Claims Right to Midland Plant Study," *Detroit Free Press*, November 7, 1985, A1. The MPSC case where Consumers Power had been told not to spend any more money on the Midland Plant was Case No. U-7850, Step 3A.
3. White, "Consumers Claims Right to Midland Plant Study," A1.
4. "Consumers to Study Completing Midland," *Herald Palladium* (St. Joseph, MI), October 12, 1985, 9.
5. Consumers Power Company, *1978 Annual Report* (Jackson, MI: Consumers Power Company, February 9, 1978, 35.
6. Consumers Power Company, *1980 Annual Report* (Jackson, MI: Consumers Power Company, February 13, 1981), 4.
7. Gina Shaw, *Building for the Future: The Story of Consumers Energy, 1886–2016* (St. Louis, MO: Essex Publishing Group, 2017), 64.
8. "Beginning Our Second Century Powering Michigan's Progress," *Consumers Power Company Publication*, 1986, A1.27.
9. Ken Kusmer, "PSI Begins Selling Marble Hill Equipment," Associated Press, October 8,1985, https://apnews.com/54b2400d19d7046de0fc13b0552ace80.
10. "Nearly Completed Nuclear Plant Will Be Converted to Burn Coal," *New York Times*, January 22, 1984, 1, 20.
11. Don Safer and Sara Barczak, "Watts Bar Unit 2, Last Old Reactor of the 20th Century: A Cautionary Tale," *Bulletin of the Atomic Scientists*, October 8, 2015, https://thebulletin.org/2015/10/watts-bar-unit-2-last-old-reactor-of-the-20th-century-a-cautionary-tale/. "Watts Bar Nuclear Plant Unit 2 Completion Project, Estimate to Complete Executive Final Report April 2012," Tennessee Valley Authority, 2012, https://tva-azr-eastus-cdn-ep-tvawcm-prd.azureedge.net/cdn-tvawcma/docs/default-source/default-document-library/site-content/news/wb2/wattsbar2_executive_etc.pdf?sfvrsn=17e5aa29_0. This TVA report should be required reading for all managers of large construction projects. In 2011, a new management team was established who took a number of steps that allowed the Watts Bar 2 project to be completed. This report details what they did.
12. See appendix 5 and table 5 for a simplified diagram and values of the Midland Plant items.
13. George White, "Consumers Board OKs Gas Plan for Midland Plant," *Detroit Free Press*, April 9, 1986, 14.

14. C. E. Gatlin Jr., Gerald C. Vellender, and James A. Mooney, "Repowering of the Midland Nuclear Station," paper presented at the Tenth Annual Industrial Energy Technology Conference, Houston, TX, September 13–15, 1988, http://hdl.handle.net/1969.1/92393.
15. Nelson E Hay, ed., *Guide to Natural Gas Cogeneration* (Lilburn, GA: The Fairmont Press, Inc., 1988), 147, 508. The term combined-cycle means that one heat source, the burning of natural gas, would be producing electricity both from the electric generators attached to the gas turbines and an electric generator attached to the steam turbine. The term "cogeneration" means that in addition to making electricity from the natural gas turbine, the heat from the gas turbine exhaust, which would be otherwise wasted, is converted to steam and used to power a steam turbine.
16. Lee Smith, "Opportunities and Challenges in Converting Midland Nuclear Site to Cogeneration," in *Opportunities and Challenges for Natural Gas* (Orlando: U.S. Department of Energy and National Association of Regulatory Utility Commissioners, February 12, 1995), 519.
17. "William T. McCormick, Jr.: CMS Energy," National Museum of American History, Smithsonian Institution, https://americanhistory.si.edu/powering/ohi/mccormk/mccormk1.htm.
18. White, "Consumers Board Oks Gas Plan for Midland Plant,'" 14.
19. Malcolm Johnson, "Consumers Woos Opinion-Makers on Midland Plan," *Lansing State Journal*, May 19, 1986, 14.
20. George White, "Consumers' Midland Plan OK'd," *Detroit Free Press*, October 23, 1986, 1.
21. William T. McCormick Jr., letter to Paul F. Oreffice, president and CEO, Dow Chemical Company, April 7, 1986. A copy of this letter was provided to the author by Jack Brown. Brown was a member of the Dow Chemical team that negotiated the solution to the Dow Chemical–Consumers Power dispute.
22. George White, "The Question Heard 'Round the State," *Detroit Free Press*, September 29, 1986, 38, 45.
23. Pete Lehman, discussion with author, February 9, 2017.
24. White, "The Question Heard 'Round the State," 45.
25. George White, "Consumers, Dow OK Midland Gas Plan," *Detroit Free Press*, January 28, 1987, 4C (and corrected *Detroit Free Press*, January 29, 1987, 3).
26. White, "Consumers Dow OK Midland Gas Plan," 33. The partners in MCV as of September 1988 were Midwest Energy Partners (a subsidiary of Asea Brown Boveri), CMS Midland (a subsidiary of CMS Energy Corp.), Coastal Midland Inc. (a subsidiary of the Coastal Corporation), CE Midland Energy, Inc. (a subsidiary of Combustion Engineering), Rofan Energy, Inc. (a subsidiary of Dow Chemical), Fluor Venture Group (a subsidiary of Flour

Corporation), and Source Cogeneration Company (a subsidiary of Panhandle Eastern Corporation).

27. See appendix 5 for a simplified diagram of plant. Gatlin, Vellender, and Mooney, "Repowering of the Midland Nuclear Station."
28. George White, "Consumers Sees End to 'Midland Problem,'" *Detroit Free Press*, October 11, 1987, 72.
29. Rodney Boulanger, interview with author, December 14, 2016.
30. John Lear, "Power of Persistence," *Detroit Free Press*, March 26, 1990, 1E, 2E. Although MCV's sales contract was reported as 806 MW increasing to 1,240 MW in Lear's article, it was not that simple. Many parties contested MCV's sales contracts for a number of years. For two examples of the cases before the Michigan Public Service Commission involving MCV and Consumers Power, see Michigan Public Service Commission Case No. U-8871 and Case No. U-10127.
31. Steam and Electric Power Agreement between the Dow Chemical Company and Midland Cogeneration Venture Limited Partnership, January 27, 1987. I have knowledge of this contract from the job I held at the Midland Cogeneration Venture Limited Partnership form 1988 to 2004.
32. Lear, "Power of Persistence," 42.
33. John Lear, "New Midland Power Plant Lauded," *Detroit Free Press*, April 10, 1990, 33.
34. "Midland Trumpets Cogeneration," *Petoskey News-Review*, April 10, 1990, 11.
35. Shaw, *Building for the Future*, 68. White, "Consumers Dow OK Midland Gas Plant," 33. "The '80s: A Look Back," *Herald-Palladium* (St. Joseph, MI), December 29, 1989, 28.
36. For a summary of some of the Michigan Public Service Commission orders and court cases, see Michigan Public Service Commission order for cases U-8871 et al and U-10127, March 31, 1993.
37. The abbreviation kp means 1,000 pounds. Laurie Valasek, email to author, January 25, 2019. Valasek was MCV's chief financial officer.
38. "MPSC approves Consumers Energy's integrated resource plan," June 7, 2019, Michigan Public Service New Release, June 7, 2019, https://www.michigan.gov.
39. "2019 Clean Energy Plan," https://www.consumersenergy.com/-/media/CE/Documents/sustainability/integrated-resource-plan-summary.ashx?la=en&hash=9F602E19FE385367FA25C66B6779532142CBD374.
40. George Bush, *Future Builders: The Story of Michigan's Consumers Power Company* (New York: McGraw Hill, 1973), 418.
41. Consumers Energy's Patricia Kessler Poppe is the daughter of William Kessler, who was a Consumers Power Company project manager for the Midland Plant. As reported earlier,

he spoke at the Midland Chamber of Commerce "Eye Opener" breakfast held on January 9, 1968, on the employment opportunities at the Midland Plant and the scheduled completion date of the Midland Plant. He was also a member of the delegation that went to Washington on October 26, 1971, to try to speed up the granting of the construction permit for the Midland Plant. She was president and CEO of CMS Energy from July 2016 to December 2020. In November 2020, it was announced that Poppe would be leaving CMS Energy and joining PG&E as CEO starting January 4, 2021.

42. Jeff Gibbs, director, and Michael Moore, executive producer, *Planet of the Humans* (Beulah, MI: Huron Mountain Films, 2019), https://planetofthehumans.com/.
43. Taylor DesOrmeau, "Southern Michigan Solar Farm Inks Power Deal with Consumers Energy," *MLive*, September 10, 2019, https://www.mlive.com/news/jackson/2019/09/southern-michigan-solar-farm-inks-power-deal-with-consumers-energy.html. This article states that this solar farm in Calhoun County, Michigan, has a maximum capacity of 149 MW and will take up almost one thousand acres.
44. Jeffrey Tomich, "Wind Turbine Shutdowns During Polar Vortex Stoke Midwest Debate," Energywire, February 27, 2019, https://energynews.us/2019/02/27/midwest/wind-turbine-shutdowns-during-polar-vortex-stoke-midwest-debate/.
45. Julian Vandal, Chris Hoch, Michael Dubois, and Casey Kopp, "Polar Vortex 2019: How Did Power Markets Weather the Storm?," *Genscape*, February 13, 2019, https://www.genscape.com/blog/polar-vortex-2019-how-did-power-markets-weather-storm.

EPILOGUE

1. Nancy Lamb, email to author, January 16, 2017. Lamb worked for Dow Chemical in public affairs. Kyle Bandlow, email to author, July 31, 2020. Bandlow was the corporate media relations director at Dow Chemical.
2. Lamb, email to author, January 16, 2017.
3. "Midland, Michigan Population 2020," World Population Review, https://worldpopulationreview.com/us-cities/midland-mi-population. According to this source, the U.S. Census estimates the 2018 Midland population to be 41,796, and the last official U.S. Census in 2010 recorded the population at 41,863.
4. Bandlow, email to author, July 31, 2020.
5. "Structures," Mid-Century Modern Midland, https://midcenturymidland.org/structures/.
6. "Canopy Walk Reaching 40 Feet High Opens," *Lansing State Journal*, October 8, 2018, A3.
7. Robert Acosta, "'I Never Thought I Would See the Day': More Than 1,000 Marched at Rally for Racial Justice in Midland," *MLive*, updated June 9, 2020, https://www.mlive.

com/news/saginaw-bay-city/2020/06/i-never-thought-i-would-see-the-day-more-than-1000-march-at-rally-for-racial-justice-in-midland.html.

8. Courtney Soule, "Meet Some of the Co-organizers of Midland's Black Lives Matter Protest," Second Wave Media Catalyst Midland, June 11, 2020, https://www.secondwavemedia.com/midland/features/BLM-Midland-061120.aspx. This article has multiple pictures of the march.
9. Dr. Etcyl Blair, telephone conversation with author, January 20, 2018.

APPENDIX 1. RADIOACTIVE WASTE CONCERNS

1. "Intelligence Reports," *Parade Magazine*, April 26, 1970, quoted in Mary Sinclair, *Nuclear Power & Public Concern*, 1968? and revised 1971, 30. Mary P. Sinclair papers 1967–1999, Bentley Historical Library, University of Michigan. On July 2, 2018 Annette Cary reported on page C4 of the *Spokesman Review* (Spokane, WA) that a nuclear storage tank at the Hanford facility dating to 1971 was taken out of service after developing a leak.
2. David E. Lilienthal, *Change, Hope, and the Bomb* (Princeton, NJ: Princeton University Press, 1963), quoted in Sinclair, *Nuclear Power & Public Concern*, 27. Lilienthal was the first chairman of the Atomic Energy Commission.
3. *Effects of Nuclear Power Plants* (Baltimore: School of Hygiene and Public Health, the Johns Hopkins University), quoted in Sinclair, *Nuclear Power & Public Concern*, 22–23.
4. Eric Sharp, "Consumers Power shows off A-plant," *Detroit Free Press*, April 21, 1979, 16C.

APPENDIX 2. NUCLEAR REACTOR TYPES

Irvin C. Bupp and Jean-Claude Derian, *Light Water: How the Nuclear Dream Dissolved* (New York: Basic Books, 1978), 49. Bupp and Derian also discuss in their book gas-graphite reactors, which were one of several additional designs besides light-water reactors. Gerard H. Clarfield and William M. Wiecek, *Nuclear America: Military and Civilian Nuclear Power in the United States, 1940–1980* (New York: Harper & Row, 1984), 1–25, 352-353. Richard A. Mueller, *Energy for Future Presidents* (New York: W.W. Norton & Company, 2012), 12–13.

APPENDIX 4. FIX FOR CRACKING IN THE WALLS OF THE AUXILIARY BUILDING

Information for this appendix is from the following sources: Jan Hamann, "Aux Underpinning Plan Adds Structural Support," *Midland Reactor,* January 1982, 1; Jan Hamann, "Freeze Wall First Support Step," *Midland Reactor,* January 1982, 3; N. M. Newmark and W. J. Hall, *Development of Criteria for Seismic Review of Selected Nuclear Power Plants, Under Contract No. AT(49-24)-0116* (Washington, D.C.: U.S. Nuclear Regulatory Commission, May 1978); Paul Rau, "NRC Sees Stipulations as Helpful to Hearings," *Midland Daily News,* July 23, 1981, 3; Paul Rau, "Underpinning N-Plant Is a Slow, Complex Task," *Midland Daily New,* September 5, 1983, 9; Paul Rau, "Digging Under an N-Plant," *Midland Daily New,* September 5, 1983, 11; Lorie Shane, "N-Plant Reinforcement Plan Presented," *Midland Daily News,* December 2, 1981, 3; Irving Spiewak and David F. Cope, *Overview Paper on Nuclear Power* (Oak Ridge, TN: Oak Ridge National Laboratory, September 1980), 36; Rodney Wieland, interview with author, May 31, 2016.

Selected Bibliography

Books

Bodansky, David. *Nuclear Energy, Principles, Practices, and Prospects.* New York: Springer-Verlag, 2004.

Brandt, E. N. *Chairman of the Board.* East Lansing: Michigan State University Press, 2003.

———. *Growth Company.* East Lansing: Michigan State University Press, 1997.

———. *We Called It Mag-Nificent: Dow Chemical and Magnesium, 1916–1998.* East Lansing: Michigan State University Press, 2013.

Bridge Magazine and the Detroit Journalism Cooperative. *The Intersection: What Detroit Has Gained, and Lost, 50 Years After the Uprisings of* 1967. Traverse City, MI: Mission Point Press, 2017.

Bupp, Irvin C., and Jean-Claude Derian. *Light Water: How the Nuclear Dream Dissolved.* New York: Basic Books, Inc., 1978.

Bush, George. *Future Builders: The Story of Michigan's Consumers Power Company.* New York: McGraw-Hill Book Company, 1973.

Campbell, John L. *Collapse of an Industry.* Ithaca, NY: Cornell University Press, 1988.

Cantonwine, Paul E. *The Never-Ending Challenge of Engineering: Admiral H. G. Rickover in His Own Words.* La Grange Park, IL: American Nuclear Society, 2013.

Clarfield, Gerard H., and William M. Wiecek. *Nuclear America: Military and Civilian Nuclear*

Power in the United States, 1940–1980. New York: Harper & Row, 1984.

Cohen, Bernard. *The Nuclear Energy Option.* New York: Plenum Press, 1990.

Cook, Constance E. *Nuclear Power and Legal Advocacy.* Lexington, MA: D.C. Heath and Company, 1980.

Denton, Sally. *The Profiteers, Bechtel and the Men Who Built the World.* New York: Simon & Schuster, 2016.

Hewlett, Richard G., and Francis Duncan. *Nuclear Navy.* Chicago: University of Chicago Press, 1974.

Jorgensen, Timothy J. *Strange Glow.* Princeton, NJ: Princeton University Press, 2016.

Levinson, Marc. *An Extraordinary Time.* New York: Basic Books, 2016.

Neer, Robert M. *Napalm: An American Biography.* Cambridge, MA: Harvard University Press, 2013.

Rockwell, Theodore. *The Rickover Effect: How One Man Made a Difference.* Annapolis, MD: United States Naval Institute Press, 1992.

Shaw, Gina. *Building for the Future: The Story of Consumers Energy, 1886–2016.* St. Louis, MO: Essex Publishing Group, 2017.

Walker, J. Samuel. *Three Mile Island: A Nuclear Crisis in Historical Perspective.* Berkeley: University of California Press, 2004.

Walker, J. Samuel, and Thomas R. Wellock. *A Short History of Nuclear Regulation 1946–2009,* NUREG/BR-0175, Rev. 2. Washington, D.C.: U.S. Nuclear Regulatory Commission, October 2010.

Ward, Geoffrey C., and Ken Burns. *The Vietnam War.* New York: Alfred A. Knopf, 2017.

Wellock, Thomas Raymond. *Critical Masses: Opposition to Nuclear Power in California, 1958–1978.* Madison: University of Wisconsin Press, 1998.

Yates, Dorothy Langdon. *Midland, Michigan: The City That Chemistry Built.* Midland: The Piney Woods Group, 2003.

———. *Salt of the Earth.* Midland: The Midland Historical Society and the Center for the Arts, 1987.

Court Cases

Calvert Cliffs' Coordinating Committee v. AEC, 449 F.2d 1109 (D.C. Cir. 1971).

Environmental Defense Fund v. Ruckelshaus, 439 F.2d 584 (D.C. Cir. 1971).

Natural Resources Council v. United States Nuclear Regulatory Commission, 547 F.2d 633 (D.C. Cir. 1976).

Nelson Aeschliman v. United State Nuclear Regulatory Commission, 547 F.2d 622 (D.C. Cir. 1976).

Power Reactor Development Co. v. International Union of Electrical, Radio and Machine Workers, 367 U.S. 396 (1961).

Vermont Yankee Nuclear Power Corporation v. Natural Resources Defense Council, Inc. and Consumers Power Company v. Aeschliman, 435 U.S. 519 (1978); 98 S. Ct. 1197 (1978).

Index

Page numbers in *italics* refer to figures and tables.

A

ABATE (American Association of Businesses Advocating Tariff Equity), 158, 162, 170, 177, 236 (n. 4)

Abraham, Karl, 124

acid rain, viii, 37

activism, 139; AEC hearings and, 17; demonstrations as "direct action," 131; environmental, vii–viii; Huron Alliance, 132; Moore and, 132, 134; "Multi-Media Confrontation" article on, 81. *See also* Sinclair, Mary

Adams, Paul L., 159

Administrative Procedure Act, 84, 86

Advisory Committee on Reactor Safeguards (ACRS), 85–86, 111, 154

AEC (Atomic Energy Commission): ASLB and, 80, 138, 154–155; *Calvert Cliffs'* lawsuit and, 9–10, 83–85; *Civilian Nuclear Power—A Report to the President—1962* by, 39; classified materials and, 57; construction problems overlooked by, 155; construction requirements of, 4, 8, 89, 97, 107; Consumers Power reactor study and, 36; creation of, 13; divided into two agencies, 214 (n. 6); EBR-1 reactor and, 47; hearings and, 17; licensing load of, 14, 15, 118; Lilienthal as chairman of, 244 (n. 2); McCone as chairman of, 47; Muntzing at, 92, 93–94; new plant reviews by, 9–10; NRC took over functions of, 118, 214 (n. 6); ORAU and, 66; public's confidence and, 14–15; quality control and, 92; reactor development and, 25; Rickover approach and, 23, 25; safety hearings held by, 202

(n. 4); safety regulations developed by, 13, 94, 107, 118; waste storage and, 186
Aeschliman, Nelson, 84, 85, 219 (n. 37)
AFL-CIO, 71
Akwesasne Notes (newspaper), 134
Alabama Power, 140, 165
Alahee, James F., 137–138
Albertson, James, 49, 50
Alden B. Dow Museum of Science and Art, 75
Allison, H. C., 8
American Association of Businesses Advocating Tariff Equity, 158, 162, 170, 177, 236 (n. 4)
American Civil Liberties Union, 133
American Electric Power, 8
American Gas Association, 169, 173
American National Standards Institute, 107
American Natural Resources Company, 173, 175
American Society for Quality Control, 149
Anders, O. U., 4
Anderson, Hugh B., 160–161, 162, 172
ANG Coal Gasification Company, 175
Ann Arbor Sun (newspaper), 93
ANR (American Natural Resources Company), 173, 175
Anton, Bob, 134
Arab oil embargo, 96, 102
Arbury, Dorothy Dow, 58
Arbury, Ned S., 68, 70, 182
Arkansas Power & Light, 52
Asea Brown Boveri, Inc., 175, 241 (n. 26)
ASLB, 80, 138, 154–155
Asperger, Robert, 31, 62, 82, 206 (n. 23)
Association of Businesses Advocating Tariff Equity, 158, 162, 170, 177, 236 (n. 4)
Atomic Bomb Casualty Commission, 14
atomic bombs, 14, 67, 76, 202 (n. 5)
Atomic Energy Act of 1954, 13, 15, 17, 66, 84
Atomic Energy Commission. *See* AEC
"Atomic Plague, The" (Burchett), 14
Atomic Power Development Association, 38
Atomic Safety and Licensing Board (ASLB), 80, 116, 117, 138, 154–155
Aymond, A. H.: as Consumers Power CEO, president, and chairman, 42, 95–96, 99, 109, 119, 167–168; Frank Kelley and, 99–100; management style of, 43, 95, 97, 99–100, 158; MPSC and, 100; resignation of, 100, 168

B

B. C. Cobb Plant, 37
Babcock & Wilcox: Kemeny Report and, 127, 228 (n. 6); PWR light-water reactors and, 39, 52, 188; Three Mile Island reactor by, 126
Bandlow, Kyle, 243 (n. 1)
Barker, Bill, 59
Barstow, Herbert D., 216 (n. 6)
BASF (Badische) Chemical Company, 31
Bay City, MI, 60, 73–74, 91; coal mines near, 28; Lone Tree Council and, 131; lumber boom in, 28; nuclear plant near, 42, 97; population of, 2; Sinclair speech in Guadalupe Center, 131
Bay City Times (newspaper), 62
Bazelon, David, 85
Bechtel, Riley P., 45
Bechtel, Stephen D., Jr. (Steve), 31; as Bechtel Group president, 45–46; at Midland Plant, 46, 209 (n. 9)

Bechtel, Stephen D., Sr., 45; cost-plus contract model of, 48, 150; Davis hired by, 47; McCone and, 47, 48
Bechtel, Warren, 45
Bechtel Company, 41
Bechtel Corp. of San Francisco, 93
Bechtel Corporation: Bechtel Jr. as president of, 31; Big Rock Point BWR and, 39, 43, 47; Consumers Power lawsuit and, 41, 98, 101; Doan and, 31; Hull as president of, 169; industrial mismanagement by, 148; Midland Plant project of, 88, 166; Midland Plant study by, 30–31; payroll of, 224 (n. 4); quality control and, 88, 89, 112, 149, 153–154; of San Francisco, 45, 93; "shit house" poet and, 104; in World War II, 47
Bechtel Group: Bechtel Jr. as president of, 46; comprised of three companies, 45–46; Consumers Power Company and, 46, 48; cost-plus contracts of, 48, 149, 150; description of, 45, 46; international companies in, 45; lawsuits brought against, 98, 101; megaprojects of, 45–48, 167; Midland Plant and, 46, 47, 48, 153; Midland Plant cancellation and, 166–167; NRC found fault with, 156; nuclear power association with, 47; nuclear projects of, 47; ownership of, 210 (n. 10); Palisades Plant and, 41, 48, 98, 101, 169; as private company, 47; profits of, 46; quality control by, 149–151; revenues of, 46; of San Francisco, 93, 45; women working for, 107
Bechtel Inc., 45
Bechtel Power Company, 128
Bechtel Power Corporation, 30, 45; Midland Plant and, 31, 150, 151, 154; Palisades Plant interest purchased by, 169
"Bechtel turn," 104
Beckrow, David, 131
Benway, Bruce, 70, 92
Beutel, Dutch, 29
Bielinski, W. Victor, 50
Big Rock Point Plant: Bechtel as prime contractor for BWR at, 39, 43, 47; Campbell on, as R&D project, 39; Consumers Power and, 41; GE equipment at, 39–40, 41, 43, 79; MW capacity of, 39, 41, 43, 208 (n. 20); problems at, 39; reactor at, 43; start date of, 39
Black employees, 78, 106, 182
Black Lives Matter, 182
Black population: Detroit, 78; Midland, 78, 182
Blair, Etcyl, 182
Blanchard, James, 164, 168, 170, 176, *176*
Blasy, Julius, 70
Blessed Sacrament Catholic Church, 55, 58
borated water tank, 116, 118
Boris, Walter R., 157–158, 162; retirement of, 167, 168; title of, 143, 236 (n. 1)
Bosscher, Harold, 31, 49, 50, 65
Boulanger, Rodney, 159, 175
Boundy, Ray, *Styrene*, 58
Boyer, Raymond F., *Styrene*, 58
Brandt, E. N. (Ned), *Growth Company*, 31
Breadfruit Rock Band, 9, *9*
breeder reactor, 16, 17, 38, 47, 188
brine: chlorine from, 28–29; wells, 28
Brinewell (newsletter), 32, 201 (n. 8)
bromine, 2, 4, 27, 82; in Bromo-Selzer, 205 (n.

8); from Midland Chemical Company, 28
Brown, Ellen, 139
Brown, Jack, 241 (n. 21)
Browns Ferry Nuclear Plant, 118
Building for the Future (Shaw), 172
Bull, Bob, 152
Burchett, Wilfred, "The Atomic Plague," 14
Burns and Rowe, 128
BWR (boiling-water reactor), 38, 39, 41, 42, 43, 47, 188

C

cadwelding, 92–93, 155
calcium, 2
calcium chloride, 32
Calvert Cliffs' Coordinating Committee, Inc, v. United States Atomic Energy Commission, 9–10, 83–85
Calvert Cliffs' Nuclear Plant, 9–10
Campbell, James H. (Jim), *32*, 35–36, 38, 144; death of, 42, 95; in *Electric World*, 40
Campbell, Susan, 173
cancer, 125, 186; leukemia, 130
carbon dioxide, viii, 177
carbon emissions, viii, 37, 177, 179
Carlson, John C., 161
Carter, Jimmy, 127
CBS Evening News (TV), 124, 147–148, 234 (n. 3), 235 (n. 8)
CE Midland Energy, Inc, 241 (n. 26)
Cederberg, E. A., 93
chain reaction, 4; of fissioning, 20, 24, 114, 187
Chapple, Simon, 71
Chatterton, Robert B., 70
chemistry, 182; Midland culture of, 3–4, 5, 76
Chernobyl reactor accident, 228 (n. 4)
Cherry, Myron, *80*; court appeals of, 87; Midland Plant construction hearings and, 69–70, 80, 81–83, 87–88; Palisades Plant and, 79, 82, 85; represented Saginaw Valley Nuclear Study Group, 80, 81, 84–85; Saginaw Interveners court appeal and, 85; Sinclair and, 60, 62, 80; temperament of, 62, 80, 81–83; U.S. Supreme Court review of Interveners case, 85–87, 111–112; Vermont Yankee case and, 85; *Wall Street Journal* article on, 80
Chicago, Illinois: AEC office in, 93; Fermi's nuclear reactor at, 20, 24; NRC office in, 152, 154
Chicago Pile-1 (nuclear reactor), 20, 187, 188
Chicago Tribune (newspaper), 135
China Syndrome, The (film), 87, 121–123, 126, 128, 135
"China syndrome" meltdown, 114, 122, 202 (n. 4)
chlorine, 4, 28
Chrysler (corporation), 236 (n. 4)
Chrysler, Walter, 3
Cincinnati Gas and Electric, 172
Circular (NRC), 115
Cisler, Walker, 36
Citizens Committee for Environmental Protection of Michigan, 60
"City of Beautiful Churches," 212 (n. 21)
civil rights, 51, 78, 131. *See also* Black employees; Midland Plant, employees; women
Civilian Nuclear Power—A Report to the President—1962 (Seaborg)(AEC), 39
Clarfield, Gerald H., 17, 202 (n. 5)

Clark, John W., 168
clay (soil), 3, 90, 116–118, 192–193, *192*
Clergy and Laymen Concerned about Vietnam (organization), 51, 76, 77
climate change, vii, viii
CMS Energy Corporation, 174, 241 (n. 26); McCormick as chairman of, 169, 176, *176*; Poppe as CEO of, 178, 242–243 (n. 41); stock dividend of, 177
CMS Midland Inc., 174, 241 (n. 26)
coal: alternative to, 8; Consumers Power source of, 168; deposits, 2–3; "dirty," 8, 12; as Dow's fuel source, 29; low-sulfur, 8, 37; Monitor Mine and, 29; in Ohio, 5, 29; prices, 5, 7, 28–29, 36, 168, 225 (n. 5); Saginaw Valley mines of, 5, 29; tons used annually by Dow, 27
coal-fired steam plants: converted from nuclear, 172; kWh cost of, 30, 39; MW capacity of, 37; pollution from, viii, 12, 31, 37–38
Coastal Corporation, 173, 175, 241 (n. 26)
Coastal Midland Inc., 241 (n. 26)
"cogeneration," 241 (n. 15)
Cohan, Leon, 160
Cohen, Bernard, 108
Columbia River, 60, 185
combined-cycle generation, 241 (n. 15)
Combustion Engineering, 39, 41, 43, 175, 188, 241 (n. 26)
concrete problem review board, 91, 92
Connell Rice & Sugar Company, 143
Consolidated National Intervenors, 62, 217 (n. 8)
construction costs, nuclear plant material/labor, *108*
construction management, 25, 62, 63, 176
Consumer Price Index, 51, 141
Consumers Energy, 177, 178, 179–180, 183
Consumers Power Company (finances): bankruptcy and, 157, 162, 167; bonds sold by, 51; cost recovery by, 161; economic growth of, 5–6; economic pressure on, 52, 96–97; expenses of, 96–97, 100, 118, 142–143, 165; finances and bank interest of, 142–144, 157–158, 162, 165, 168; first mortgage bonds of, 97, 142–143, 144; inflation and, 51–52; as investor-owned, 5, 52, 145, 157, 161, 166, 167; MPSC and costs of, 33, 142, 158, 161, 163; profit for, 38; rate increases and, 98–100, 158, 160–163, 166, 171; stagflation and, 97; stocks and dividends in, 144, 145, 157, 167
Consumers Power Company (management): Aymond as CEO, president, and chairman of, 42, 95–96, 99–100, 109, 119, 167–168; Boris as V.P. in, 143, 157–158, 162, 167, 168, 236 (n. 1); Campbell as officer of, 32–33, *32*, 35–36, 38, 40, 42, 95, 144; Cook as V.P. in, 172, 191, 193; employment at, 7; Griswold as treasurer in, 143; Keeley as QA director in, 117; Kessler and, 49–50, 71, 103, 178, 242 (n. 41); McCormick replaced Selby at, 168–169, 171–176, *176*, 177, *197*; Peck hired by, 90, 91, 94, 104, 141, 152; Selby as president of, 42–43, 68, 95–102, 109, 119, 144, 157–158, 168–167
Consumers Power Company (utility): AEC nuclear reactor study and, 36; annual report of, 95; apology to Midland workers by, 108; Bechtel bought Palisades plant interest from, 169; Big Rock Plant and,

39–40, 41, 43, 79; as conservative, 35; as Consumers Energy, 177, 179; Cook's Midland Options Study in, 172–173; Detroit Edison and, 36; Dow Chemical purchase agreement and, 32, *32*, 101–102, 219 (n. 37); Dow Chemical steam from, 5, 137–138; Dow contract termination and, 138, 157; electricity futures, 179; electricity supply plan failure and, 1; future plan for 2040 of, 179; growth forecasts by, 144, 160; industrial mismanagement by, 94, 100, 148, 156; *Lansing State Journal* on Midland Plant and, 48, 155; lawsuits and, 101, 111, 138, 158, 170, 174; Marysville synthetic gas plant and, 98–100; MCV MW contract with, viii, 177; Midland nuclear reactors and, 5, 7; Midland Plant cancellation and, 162–163, 165; Midland Plant completion by, 95, 98, 111, 138, 162, *163*; Midland Plant failure and, vii–viii, 137, 165; Midland Plant repairs and, 118, 137, 140, 155, 191, 193; NRC and, 118, 138–140, 147, 152, 155–156, 191; Oakway IV repurchase and, 143; origins of, 36; Palisades Plant of, 40, 101; Plateau Resources Limited and uranium for, 111; pollution control and, 37; power as future of, 8, 19, 35–36, 38, 144; projects cancelled by, 97, 100; reactor manufacturers used by, 41; Rickover management style and, 25; sinking building fix by NRC, 115; steam production by, 5, 19; turnkey contracts and, 40, 41; in uranium business, 111–112; water-cooled reactor choice of, 19. *See also* CMS Energy Corporation; quality assurance; quality control

Consumers Power Company annual reports, 237 (n. 11); of 1967, 5, 35; of 1968, 52; of 1969, 37; of 1971, 70; of 1974, 94–95; of 1975, 100; of 1977, 111; of 1980, 233 (n. 22); of 1981, 142, 233 (n. 22); of 1983, 142–143, 145, 233 (n. 22), 237 (n. 11); of 1984, 233 (n. 22), 237 (n. 11)

Consumers Power–Dow Chemical contract, 138, 157, 170, 177

Consumers Power Finance, 143

containment buildings, 114, 122; Midland Plant, 104, 116, 189, 191, 193; Three Mile Island, 123

control rods, 122, 187

Cook, James, 172, 191, 193

Cooke, Thomas C., 103

Cordiner, Ralph, 38

cost-plus contract model, 48, 149, 150

Coultrip, Phil, 9, *9*

COVID-19 pandemic, 182

Crew, Dana, 107

Cronkite, Walter, 124, 128

Crow, David, 235 (n. 8)

Curry, Don, 125

D

Daily Express (newspaper), "The Atomic Plague," 14

dams: Bechtel built, 45, 46; Consumers Power Michigan river, 36–37; hydroelectric, 5; MW generation and, 37; site flooding and, 116, 182

Dan E. Karn Generating Plant, 102

Davis, Leo, 46

Davis, W. Kenneth, 47

Davis, Will, 25
"Declaration of Nuclear Resistance," 132
demonstrations: anti-nuclear "direct action," 131–132, *133*; Black Lives Matter, 182; Moore and, 132, *134, 135*; napalm and, 76; Sinclair and, 132
Detroit, MI, 3; civil disruption and racism in, 78; Fermi 1 plant site and, 15, 16, 38; labor contracts and, 16; power for auto industry and, 21; protesters from, 135
Detroit Edison Company (utility): Cisler as president of, 36; Consumers Power and, 36; cost recovery by, 161; expenses of, 100; Fermi 2 plant built by, 42, 142; finances of, 142; GE BWR used by, 42; growth forecasts by, 160; industrial kWh rates of, 237 (n. 12); McCarthy as chairman of, 168; nuclear plant plans of, 42; power plant management by, 25; PRDC and, 15, 38, 161
Detroit Free Press (newspaper): on AEC and plant safety, 94; on Aymond and rate increases, 99; on Aymond's reputation, 100; diagram of Midland nuclear steam making process in, *190*; diagram of Midland Plant gas conversion in, *196*; on Midland Plant radioactive waste, 186; on Midland protest march, 135; Mooney on plant repair in, 140; on radiation effects in Japan, 14; Selby interview in, 101
deuterium, 188
DeWitt, David Scott, 174
Doan, Herbert D. (Ted), 216 (n. 6); as Dow Chemical president, 4, 31, *32*
Doan, Junia, 216 (n. 6)
Doan, Leland I. (Lee), 31
Dorman, Linneaus, 78
Dorman, Phae, 78
Douglas, Michael, 122
Dow, Alden B., 74, 75
Dow, Barbara C., 216 (n. 6)
Dow, Dorothy, 58
Dow, Grace A., 74–75, *75*
Dow, Helen (Mrs. Macauley Whiting), 75, 216 (n. 6)
Dow, Herbert Henry (founder), 76, 183, 216 (n. 6), 238 (n. 7); as company founder, 3, 27, 74; daughter Dorothy and, 58; granddaughter Helen and, 75, 216 (n. 6); son Alden B. and, 74, 75; son Willard and, 29, 74
Dow, Herbert H., II (grandson), 216 (n. 6)
Dow, Vada, 75
Dow, Willard, 29, 74
Dow Botanical Gardens, 75
Dow Chemical Company, 188–189, 241–242 (n. 26); attorney Brown and, 241 (n. 21); attorney Nute of, 102, 137; Black employees in, 78, 106, 182; bleach plant of, 29; *Brinewell* newsletter of, 32, 201 (n. 8); coal used by, 5, 27, 29; competitive edge needed by, 65; compounds used by, 1–2; Consumers Power contract with, 138, 170, 177; Consumers Power purchase agreement and, 32, *32*, 101–102, 219 (n. 37); Consumers Power sued by, 117, 170, 174; Doan as president of, 4, 31, *32*; employees, 4, 7, 30, 74, 78, 106, 181–182, 243 (n. 1); formation of, 28–29; foundations associated with pioneers of, 215–216 (n. 6); Freeport plant of, 30, 69; Gerstacker as chairman of, 31,

75; government needs and, 77; MCV agreement and, 175–176, 242 (n. 31); merged with DuPont, 181; Merszei as president of, 30; Michigan Division, 174; in Midland complex, 4, 30, 74; Midland Division, 31–32, 49, 65, 201 (n. 8); Midland Plant steam used by, 7; Midland's first reactor and, 4–5; Monitor Mine and, 29; napalm made in Midland by, 51, 76–77, 216 (nn. 11–12); neutron source for, 4; Nuclear Energy Project at, 71; pollution from, 30; products made by, 4, 30; production costs of, 30, 69; profit point of, 35; racial equality in, 78, 106, 182; sales numbers of, 30; Sinclair questioned actions of, 53; size of, 53; Whiting as director of, 29, 65, 71, 75; World War I and, 4, 73, 74; World War II and, 4, 73

Dow Chemical-Consumers Power Company purchase agreement, *32*, 33, 101–102, 219 (n. 37)

Dow Chemical U.S.A., 158

Dow Corning Corporation, 58, 104, 177

Dow family philanthropy, 74, *75*, 75–76, 215–216 (n. 6)

Dow Inc.: employees of, 181, 182; MCV plant steam for, *196*; Midland research nuclear reactor of, 181

Dow Male Chorus, 74

Dow Music Building, 74

Dow USA, 65, 138, 174

DowDuPont, 181, 182

Dowflake, 32

Dresden project, 47

Duke Energy (utility), 25

Duke Power Company, 52

DuPont de Nemours & Company, 181, 182

E

EBR-1 reactor, 47

Ecology Law Review, 81

Eisenhower, Dwight, 47

Electric World (journal), 40

electricity: demand decline and, 100; in kWe, 108; kWh and, 39, 112, 237 (n. 12); low-cost, vi; rising costs and, 100. *See also* Consumers Power Company (utility)

emergency core cooling systems (ECCS), 15, 71, 83–84, 202 (n. 4)

emergency generator building, 113, *113*, 114–115, 116, 153

Emerson Park, 133

Engineering News-Record (journal), 46, 51

Engler, John, 164, 176

Enrico Fermi Atomic Power Plant. *See* Fermi 1

environmental activists, vii–viii, 17, 60, 62, 67, 97; Asperger as, on bond ratings, 82; Cherry and, 80, 82, 85; circuit court and, 85; Like, on agency proceedings, 81

Environmental Defense Fund v. Ruckelshaus, 85

Environmental Protection Administration, 107

Environmental Protection Agency, 71

Erie, Lake, 42

F

Fabian, Blaine, 124

Farley Nuclear Plant, 140

Farrar, Bud, 67

Fauver, Dorothy, 78, 217 (n. 20)

Fermi, Enrico, 20–21, 24, 187
Fermi 1 (nuclear power plant): breeder design of, 16; coolant in, 16, 19, 188; location of, 15, 16, 38; meltdown in, 1, 16–17; *Power Reactor Development Co. v. Electricians*, 16; PRDC built, 15, 38; Reuther opposed, 16, *16*
Fermi 2 (power plant), 42, 142
Fermi 3 Plant (nuclear power plant), cancellation of, 100, 162
Fieser, Louis, 76
fill dirt: compaction and, 116; as foundation, 115–116, 118, 191, 192, *192*, 193; on Midland plant floodplain, 115–116
first mortgage bonds, 97, 142–143, 144
First United Methodist Church, 66
Fischer, Roger, 162, 164, 171–172
fission: chain reaction of, 4, 20, 24, 114, 187; Chicago Pile-1 reactor and, 187; control rods and, 122, 187; environment and, 60; Fermi's first controlled, 20–21, 24, 1987; fissioning process and, 20, 187; industrial use of, 36; materials, 36; at Midland Plant, 4, 21; radioactive materials from, 123; in Shippingport reactor, 24; uranium fuel for 4, 86–87, 111–112, 144, 186, 187, 225 (n. 5)
fixed-price contracts, 48, 52
Flint, MI, 2, 91; civil disruption in, 78; unemployment in, 141
Flint Voice (newspaper), 132
Florida Power & Light, 48
Florida Power Corporation, 52
Fluor Corporation, 175, 242 (n. 26)
Fonda, Jane, 51, 122
Foote, W. F., 36
Ford, Gerald, 169
Ford, Henry, 3
Ford Motor Company, 21, 236 (n. 4)
foundations (philanthropic), 74, *75*, 182, 215–216 (n. 6)
Freeland, MI, 68, 139
Freeport, Texas, 30, 69, 138
freeze wall, 192
fuel: bromine and, 28; chemical, 187; coal, 28, 29, 30, 37, 40, 187; cost of, 8, 28–30, 38–39; fossil, vii, viii, 29; fuel cell and, 43; natural gas, viii, 29–30, 187; non-carbon, viii; nuclear sourced, 8, 38–40, 188; oil, 29; pollution for, 30, 37; solar, viii, 8, 43, 177–179, 243 (n. 43); steam, 30, 37; wind, viii, 177–179
fuel rods, 16, 23
Fukushima Daiichi Power Station (nuclear), meltdown at, 114–115, 204 (n. 22), 226 (n. 16)
fusion, nuclear, 60

G

Gallandt, Chris, 106
Garde, Billie, 148, 156, 235 (n. 8)
gas-cooled reactor, 47
gas-graphite reactors, 244 (appendix 2)
gas turbines, viii, 174, 176, 195, *196–197*, 241 (n. 15)
General Electric (GE): Big Rock Point and, 41; Cordiner of, 38; Detroit Edison BWR and, 42; light-water BWRs made by, 38–40, 188; MW capacity of BWR from, 39; orders lost by, 52; Selby worked at, 100–101; turnkey contracts and, 40
General Motors, 161, 168, 170, 236 (n. 4)

geology: Chernobyl reactor and, 228 (n. 4); clay, 3, 90, 116, 117, 192; fill dirt, 116, 117, 118, 192, *192*, 193; glacial till, 90, 116, 118, 192; glaciation and, 2–3, 5; gravel and, 90; Lower Peninsula sinking and, 3, 113–115, 118–119, 129, 137–139; Michigan Basin, 1, 3; sand, 3, 90, 117–118
Gerstacker, Carl A., 31, 75
Gerstacker, Rollin M., 216 (n. 6)
Gibbs, Jeff, *Planet of the Humans*, 178, 243 (n. 42)
glacial clay, 3, 116, 118
glacial till, 90, 116, 118, 192
glaciation, 2–3, 5
Government Accountability Project (GAP), 148, 151–152, 156, 235 (n. 8)
Grand Rapids, MI, 37, 135
"Great Bandwagon Market," 8
Great Lakes, solar and wind power, 178–179
Great Lakes Energy Alliance, 132–133
Great Plains plant, 175
Greenwood Plant, cancellation of, 100, 162
Griffin, Robert, 9, *10*
Griswold, Richard, 143
Griswold, Thomas, Jr., 27
Groeschen, Jack, 105
Gronski, Mike, 166, 238 (n. 7)
Growth Company (Brandt), 31
Guadalupe Center, Bay City, 131

H

H. H. Dow High School, 9, 12
Hallam plant, 47
Hanford nuclear facility, 60, 244 (n. 1); buried radioactive waste at, 185
Harrington, John, 126
Harrisburg, Pennsylvania, 123, 134
Hayes, Michael D. (Mike), 136–137, 165–166, 168
"heavy water," 47, 188
Heins, Gordon, 81
Henry, Hunter W., 158, 174
Herald-Palladium (newspaper), 151–152
Herbert D. and Junia Doan Foundation, 216 (n. 6)
Herbert H. and Barbara C. Dow Foundation, 216 (n. 6)
Herbert H. and Grace A. Dow Foundation, 74–75, *75*, 182
Higginbotham, Adam, 228 (n. 4)
Hiroshima, Japan, 14, 67, 76, 202 (n. 5)
Hughes, Sara, 130–131, 235 (n. 8)
Hull, Cordell, 169
Humboldt Bay project, 47
Huron, Lake, 3, 60, 227 (n. 26); nuclear plant at, 42
Huron Alliance, 132
hydrogen, 58, 114–115, 124, 130, 188, 204 (n. 22)
hydrogen bomb, 58, 130

I

Idaho, EBR-1 in, 47
Ieuter, Cal, 170
inflation: Consumers Power Midland Plant and, 52; economy rise and, 97; interest charges and, 141; Vietnam War caused, 50–51, 97, 210 (n. 7)
International Brotherhood of Electrical Workers, Local 58, 150
International Union of Electrical Workers, 16
intervenors: Mapleton, 80–81, 84, 139, 213 (n.

29); Midland Plant hearings and, 15, 17, 69, 70, 81, 128, 139; national, 62, 217 (n. 8); Palisades Plant, 69; Saginaw, 60, 61–62, 80, 83, 85; Shippingport and, 24; Sinclair as, 131–132, 139, 213 (n. 29); soils hearings and, 139; Stamiris as, 139
investor-owned utilities, 5, 17, 52, 145, 157, 161

J

J. H. Campbell Plant, 37
Jackson, MI: Consumers Power Company in, 36, 101, 165, 167; MCV offices and, 175
Jackson Citizen Patriot (newspaper), 167
Jefferson Intermediate School, 82
Jersey Central Power & Light Company, 7, 39, 52
Johnson, Lyndon B., 5, 50–51, 210 (n. 7)
Joint Committee on Atomic Energy, 24, 71

K

Kasuboski, Wally, 134
Kay-Simons, Edna, 131
Keeley, Gilbert S., 117
Kelley, Frank: as attorney general, 158–160; Aymond and, 99–100; as Consumers Power critic, 99, 158–159, 160–162, 166, 177; Coy and, 233 (n. 28)
Kemeny, John G., 127
Kemeny Report, 127, 228 (n. 6)
Kendell, James A., 70
Kennedy, John F., 50; administration of, 39
Keppler, James G., 87, 93, 94, 148, 154–156
Kern, Edward, 67
Kerr, William, 60
Kerr-McGee plant, 134, 231 (n. 26)
Kessler, William, 49–50, 71, 103, 178, 242 (n. 41)
Kids Say the Darndest Things! (Linkletter), 10
kilowatt (KW): Consumers Power sales of, 112; as electricity, *108*; Midland Plant output as, 7; sales of, 112, 138; Shippingport output as, 39
Kimball, Jeffrey, 139
Knopick, Paul, 169
Kornegay, Jeri, 167
kWe (kilowatt of electricity), *108*
kWh (kilowatt hour), 39, 112, 237 (n. 12)

L

labor unions, 9; AFL-CIO as, 71; and construction costs, *108*; and contracts, 16, 106; equal opportunity employment and, 104–106; Union Local 85, 105; and politics, 174; UAW as, 15–17, *16*, 58, 133
Lamb, Nancy, 243 (n. 1)
Lansing State Journal (newspaper), 48, 155
Lehman, F. R. (Pete), 174
Lein, Linda (Sims), 104–105, *105*
LeMay, Curtis, 76
Lemmon, Jack, 122
leukemia, 130
Levin, Carl, 176
"License for a Nuclear Power Plant" (N-rally sign), 10, *11*, 70
"light water," compared to "heavy water," 188
light-water boiling water reactor (BWR), 188
light-water pressurized reactor (PWR), 23, 41, 42, 43, 52, 188; Fermi 1 design replaced by, 38
Like, Irving, 81
Lilienthal, David, 244 (n. 2)
Linkletter, Art, *10*; *Kids Say the Darndest Things!*, 10, 201 (n. 14)

Lone Tree Council, 238 (n. 7); demonstration by, 132; Midland Plant soils criticized by, 139; name origin, 131; Sinclair and, 132; soils hearings and, 139; Terry Miller and, 57, 131, 139; Warner as president of, 139, 148
lone tree legend (Native American), 131, 238 (n. 7)
Lowenstein, Robert, 71, 72

M

Macauley and Helen Dow Whiting Foundation, 75, 216 (n. 6)
Maddox, James F., 71
magnesium, 2, 4
Mann, Joseph, 166
Mapleton, MI, 60, *61*, 80–81, 139, 213 (n. 29)
Mapleton Intervenors, 80–81, 84, 139, 213 (n. 29)
Marabito, Russ, 152
Marble Hill Nuclear Plant, 172
Marshall, Wendell, 139, 213 (n. 29)
Marshall, William, 71
Marysville synthetic gas plant, 98–100
Massachusetts Institute of Technology, 36, 168
McCarthy, Walter J., Jr., 168
McCone, John Alex, 47, 48
McCormick, William T., Jr. (Bill), 168–169, 171–175, *176*, 177, *197*
McCue, Bob, 164
McLogan, Matthew E., 167
MCV. *See* Midland Cogeneration Venture Limited Partnership
MCV Plant: converted nuclear plant site and, viii, 173–178; dedication of, 176, *176*; gas turbines of, 176, 195, *196–197*
megawatts (MW): costs and, 179; of Hardy Dam, 37; of nuclear plants, 20–21, 24, 37, 39–43, 178, 195, 208 (n. 20), 242 (n. 30); solar generation of, 243 (n. 43); wind farms and, 179
Meier, Jerry, 9, *9*
meltdown: AEC guidelines and, 83; called "China syndrome," 114, 122; causes of, 15, 83, 114, 202 (n. 4); Chernobyl reactor accident and, 228 (n. 4); explosive dangers of, 114; in Fermi 1, 16; in film *The China Syndrome*, 87, 121–123, 126, 128, 135; at Fukushima Daiichi Power Station, 114–115, 204 (n. 22), 226 (n. 16)
Merszei, Zoltan, 30
Mesler, Marge, 30
Metropolitan Edison Company (utility), 52, 124–126, 127, 128
Michigan: Coy as assistant attorney general of, 233 (n. 28); energy demand in, 40, 112; energy production in, 5; geology of, 1–3, 5; industries in, 3; Kelley as attorney general of, 99–100, 158–162, 166, 177, 233 (n. 28); 1960s economy prospered in, 5–6; as nineteenth-century lumbering state, 27–28; recession in, 141; rivers and potential for energy, 36–37; salt in, 28; unemployment in, 141. *See also* geology
Michigan, Lake: Big Rock Point plant at, 39; Palisades Plant at, 40, 79
Michigan Basin, 1
Michigan Citizens Lobby, 158, 162
Michigan Consolidated (utility), 99
Michigan Financial Crisis Council, 168
Michigan House of Representatives, 136, 165,

174
Michigan Public Service Commission. *See* MPSC
Michigan Senate, 164, 174, 176
Michigan Supreme Court, 142, 159
mid-century modern design, in Midland, 73–74, 182
Mid-Century Modern Midland, 182
Midland, MI, *2*; Arbury park in, 182; architectural heritage in, 73–74, 182; "Bechtel turn" lane in streets of, 104; Black Lives Matter march in, 182; Black population and, 78, 182; bromine brine wells in, 28; bromine fog in, 82; business sites in, 4, *61*; chemistry culture and natural resources of, 3–4, 5, 76, 182; *The China Syndrome* film shown in, 126; as city of churches, 55, 58–59, 66, *66*, 73, 74, 76; as "City of Modern Explorers," 182; civil disruption in, 78, 182; clergy supported DOW in, 51, 76–77; Consumers Power nuclear plant welcomed by, 7, 8, 67, 73; COVID-19 in, 182; Dow Chemical complex in, 4, 74; Dow employees in, 30, 74, 182; Dow foundation grants to, 74, *75*, 182; Dow Inc. located in, 181–183, *196*; Emerson Park protest in, 133–134; employment in, 67; energy for, 67; floods in, 182; Herbert Dow changed, 3; high school in, 3, 9, 12, 57, 130, 182; innovation in, 182; Jefferson Intermediate School hearings at, 82; Mann as mayor of, 166; MCV offices in, 175, 182–183; as nineteenth-century lumber town, 27–28, 73, 76; pollution in, 12, 30, 177; population of, 7, 8, 78, 182, 243 (n. 3); Poseyville Road in, 4, 75, 134; power plant in, vii, 4–5, 67, 89, 166, *176*; property taxes and, 166; protest marches in, 130, 133–135, *133*, *134*, *135*, 182; real estate market in, 166; Whiting Forest in, 182; Whiting Overlook Park in, 4, 75, 133–134, *134*; workforce growth and culture in, 104–107. *See also* MCV Plant; Midland Plant, nuclear; Speak Up Now rally; Tittabawassee River
Midland Center for the Arts (MCFTA), 75, *75*
Midland Chamber of Commerce, 8–9, 12, 49, 65, 70, 93, 107, 170, 212 (n. 21), 243 (n. 41)
Midland Chemical Company, 28
Midland Clergy Association, 76, 77
Midland Cogeneration Venture Limited Partnership (MCV), viii, *176*, 183, *196*, *197*; Boulanger as president of, 159, 175; commercial electricity generation of, 176; Consumers Power agreements with, 175, 242 (n. 31); converted site of nuclear plant and, viii, 173–178; Dow Chemical power agreement with, 175–176, 242 (n. 31); economic impacts of, 177, 182; employment by, 177; Mooney and, 238 (n. 2); MW capacity of, 177, 178; partners in, 241 (n. 26); sales contracts of, 176, 177, 242 (n. 30); success of, 177; Valasek as CFO of, 242 (n. 37)
Midland Community Center, 49, 74, *75*
Midland County Historical Society, 59, 212 (n. 21), 217 (n. 20)
Midland Daily News (newspaper): on Babcock & Wilcox reactors, 52; on "Challenge Given to All Residents to Attend N-Rally," 8; on "Chemical Center"

Midland, 4; Doan on Plant site in, 31; Dow's first nuclear reactor photo in, 4; on EPA and Plant status, 71; interveners and soils hearings in, 139; letters on Kelley and rates in, 161; "Many Yanks Being pressed into Second Vietnam Tours" in, 50; on Midland Plant costs, 225 (n. 5), 140; on Midland Plant site flooding, 153; on MNPC plant support, 68; on napalm, 216 (nn. 11–12); on NRC, 89; on nuclear protests, 131, 133, 135; Overberger on Plant license in, 12; Rumple as editor of, 56, 57; Sinclair on nuclear safety to, 56–57, 136; on Three Mile Island accident, 124–128; on Vietnam War, 50–52

Midland Library, 74

Midland Music Foundation, 74

Midland Nuclear Power Committee. *See* MNPC

Midland Options Study, 172–173

Midland Plant, construction: AEC delayed, 8, 13, 17, 89, 90–91; Cherry and permit for, 69–70, 79, *80*, 80–84, 87–88; concrete problem review board and, 91, 92; Construction Completion Program at, 153, 155; construction delays at, 83, 87–88, 90–91, 97, 111, 162, 165; construction of, vii–viii, 7, 25, 72, 89–91, *90*, 102–104, 116–118, 153, 243 (n. 41); construction permit for, 8, 12–13, 15, 17, *80*, 69–72, 81–82, 90–92, 111–112, 243 (n. 41); Consumers Power and repairs at, 118, 137, 140, 155, 191, 193; costs to complete, viii, 25, 98, 107, 111, 140–141, 148, 162, *163*; earthquake criteria and, 193; flooding at, 153; permit hearings and, 71; site of, *6*, 60, *61*, *90*, 115; "soil problems" and, 139, 158; soils hearings and, 139

Midland Plant, employees: apprentice programs and, 105, 106, 150, 166; Black workers in, 106; employee drug problem at, 152; employee workforce at, 102, 103–106, 108–109, 243 (n. 41); engineer Wheeler at, 94, 118, 119; layoffs at, 90–91, 153; Peck assigned to, 90, 91, 94, 104; "shit house" poet at, 104; women working in, 104–107

Midland plant, finances: coal cost and, 37; cost projections for, 37, 52, 72, 98, *98*, 112, 162, *197*, 209 (n. 9); costs to operate, 30; finance problems of, 141–142, 162

Midland Plant, nuclear: acid rain and, viii; Bechtel as contractor for, 43, 46–48, 93, 94, 149; Bechtel study for dual purpose of, 30–31; benefits of, 30, 89; bureaucratic maze of, 8; bromine fog and, 82; cancellation of, viii, 162–163, 165–167, 169; cancelled plant Options Study, 172–173; Consumers Power apology at, 108; Consumers Power involvement in, 93–94, 107, 111, 118, 162, 165; Consumers Power proposals, viii, 7, 12, 30, 83, 177, 191, 193; gas-fired conversion diagram in, *196*; in-service date delays at, 112; intervenors at hearings for, 15, 17, 69, 70, 81, 128, 139; Kemeny Report and, 127; Kessler and, 49–50, 71, 103, 178, 242 (n. 41); KW output of, 7; MW capacity of, 20, 21, 37, 41, 43; NRC and, 117–118, 191; NRC safety review of, 89, 155; NRC violations by, 147–148, 153, 156; protest marches against, *133, 134*; quality assurance at, 92, 94, 112, 155; quality control at, 89–91, 92–94, 112, 119, 147–151; QIP at, 152–153; radiation and, 130; regulatory updates and, 107, 108; safety of, vii, 90–91; Selby and, 102, 127–128, 145, 162–165, 167; steam process at, 115, 173, 175–176, 189, *190*, 195; steam produced by, 7; steam requirements of, 32; success of, 50; Supreme Court and, 111–112. *See also* Midland Cogeneration Venture Limited Partnership; Speak Up Now rally

Midland Plant, physical site: auxiliary building, 89–90, *90*, 108, 116–117, 186; auxiliary building repairs, 118, 138–141, 162, 191–193, *192*; auxiliary building wall crack and settling, 118, 138–140, 153, 162; "ballroom" excavation of, 193; borated water tank repairs in, 116, 118; brothel graffiti in, 104; cadwelding problems at, 92–93, 155; clay beneath, 90, 193; concrete at, 91–92, 116; containment building at, 104, 113, 116, 189, 191, 193; control rods in, 122, 187; control room in, 117, 191; control tower and, 192; cooling pond at, 89–90, *90*, 115, 117; emergency core cooling systems (ECCS), 15, 71, 83–84, 202 (n. 4); emergency generator building at, 113, *113*, 114–115, 116, 153; fill-dirt foundation and, 115–116, 118, 191, 192, *192*, 193; freeze wall of, 192; gas turbines in, viii, 174, 176, 195, *196–197*; geology and, 1; MCV turbines at, 176, 195, *196–197*; as natural gas combined-cycle cogeneration plant, 173; nuclear fuel at, 191; office buildings at, 173; outage building at, 153; power block area in, 89, *90*; PWR at, *190*; radioactive waste storage at, 87, 185–186, 219 (n. 37), 227 (n. 22); reactor containment silo structures at, *113, 192, 197*; reactors at, 41, 43, 91, 111; service water intake structure at, 116, 117, 118; sinking buildings and, 113–119, 129, 137–139, *197*; site converted to MCV gas-fired plant, 173, 176–178, *176*, 195, *196–197*; steam line at, 173, 175, 191; steam turbine building at, 89, 90, *90, 113*, 114–115, 143, 192–193, *192, 197*; switchyard, 173; "tendon sheath" problems at, 112, 155, 225 (n. 6); Tittabawassee River site of, *6*, 60, *61, 90*, 115; tunnel under, 118, 192, *192*; "turbine roll," at, 163–164; Unit 1 reactor at, 111, 112; Unit 1 turbine at, 90, 111, 143; Unit 2 turbine at, 111, 143; wall cracking at, 117–118, 191; warehouse facilities at, 173; water table beneath, 192

Midland Reactor (newspaper), 104, *105*, 105–108, 193, 224 (n. 10), 245 (appendix 4)

Midland Republicans, 60

Midland Symphony, 55, 57, 74

Midwest Energy Partners, 241 (n. 26)

Miller, Dick, 11–12

Miller, Don, 152

Miller, Gary, 123, 229 (n. 13)

Miller, Richard T., 70

Miller, Terry, 57, 131, 139

MNPC (Midland Nuclear Power Committee): chairman of, *66*; Congress and, 93; ORAU workshops and, 68; permit for Midland Plant and, 65; Sinclair opposed by, 68–69, 121; Speak Up Now rally and, 8, 9

Moll, Connie, 62
Mooney, James Arthur (Jim), 140, 165, 238 (n. 2)
Moore, Michael: activism of, 132–134, *134*, 178; at *Flint Voice*, 132; *Planet of the Humans*, 178, 243 (n. 42); *Roger and Me*, 132
Moss, Sally, 58
Mother Jones (magazine), 47
MPSC (Michigan Public Service Commission), 242 (n. 30); Aymond and, 100; Boris and, 167; Consumers Energy purchases from MCV, 183; Consumers Power rate increase and, 171; Marysville Plant and, 98–100; Midland Plant conversion and, 173–174; Midland Plant costs and, 163; Selby and, 119, 167; utility cost regulation by, 33, 142, 158, 161
Mudd, Pegg (Woodby), 107
"Multi-Media Confrontation: The Environmentalists' Strategy for a No-Win Agency Proceeding," 81
Muntzing, L. Manning, 92, 93–94
Murphy, Patricia, 131
Muskegon County, MI, 36–37
MW. *See* megawatts
MWh (megawatts per hour), 179

N

N-Rally. *See* Speak Up Now rally
Nader, Ralph, 62, 129
Nagasaki, Japan, 14, 76, 202 (n. 5)
napalm, 51, 76–77, 216 (nn. 11–12)
National Environmental Protection Act, 83–85, 86
Native American activism, 57, 131–132, 134, 139, 148, 238 (n. 7)
natural gas, viii, 173
natural gas combined-cycle cogeneration plant, 173
Natural Resources Defense Council v. U.S. Nuclear Regulatory Commission, 219 (n. 37)
Nautilus (SSN-571) (submarine), 19, 23
Ned S. Arbury pocket park, 182
NEPA (National Environmental Protection Act), 83–85, 86
neutrons, 4, 19, 39, 68, 116, 187
New York Times, 77
Newgen, Joel, 104
Nine Mile Point plant, 106
Nixon, Richard, 10, 83, 84, 95, 96
North, Wayne, *66*, 68, 71, 73, 76, 92; accident potential and, 121–122, 214 (n. 10), 228 (n. 1); as Midland Plant supporter, 67; on MNPC, 66; Sinclair debate and, 69, 121
Northern Michigan Exploration Company, 175
NRC. *See* U.S. Nuclear Regulatory Commission
"Nuclear Environmentalism—A Case Study" (Sinclair), 212 (n. 23)
nuclear fallout, 14, 228 (n. 4)
Nuclear Fuel Services plant, 47
nuclear fusion, 60
Nuclear Power, Benefits and Risks (TV), 56
nuclear power, World War II bomb use of, 14, 67, 76, 202 (n. 5)
Nuclear Power & Public Concern (Sinclair) (booklet), 56, 58, 60, 185–186
nuclear power plants: cancellations, viii, 100, 162–163, 165–166, 172–173; for civilian purposes, 7, 13, 19; construction cost

increases and, 25, *108*; conversion to coal-fired, 172; conversion to natural gas, 173; electricity first produced by, 7; illness and, 130; MW capacity of, 20–21, 24, 37, 39–43, 178, 195, 208 (n. 20), 242 (n. 30); MW power and, 40, 175–176, 177; output of, viii; true cost of, 40; turnkey loss-leader contracts and, 40, 41; utility order growth for, 7–8, 40. *See also* radiation

nuclear reactors: atomic bomb, 187; boiling water (BWR), 38, 39, 41, 42, 43, 188; breeder, 16, 17, 38, 47, 188; chain reaction in, 187; Chernobyl, 228 (n. 4); Chicago Pile-1, 20, 187, 188; civilian use and, 19; commercial, 23; control rods in, 122, 187; coolant in, 16, 19, 83, 150, 202 (n. 4); core meltdown in, 83, 114, 202 (n. 4); EBR-1r, 47; ECCS in, 71, 83–84, 202 (n. 4); Fermi 1 plant, 38, 188; fissioning process in, 187; fuel for, 188; gas-cooled, 47; gas-graphite, 244 (appendix 2); heavy-water, 47, 188; light-water, 23, 38, 39, 41, 42, 43, 188, 202 (n. 4); Midland's first, 4; pressurized water, 24, 25, 127, *190*; pressurized/PWR light water, 23, 38, 41, 42, 43, 52, 188; radiation produced by, 187; safety of, 15, 19

Nuclear Regulatory Commission (NRC), 83; AEC functions given to, 118, 214 (n. 6); attorneys for, 139; Bechtel relationship with, 156; on Big Rock Plant, 208 (n. 20); *Circular*, 115; Consumers Power relationship with, 115, 118, 138–140, 147, 152, 155–156, 191; design specifications of, 107–108, 193; Dow Chemical negotiations with, 137; history of, 118; inspection standards of, 87, 88–89, 94, 147, 148, 154; jurisdiction of, 117, 127, 214 (n. 6); Kemeny Report criticized, 127; Keppler at, 87, 93, 94, 148, 154–156; Licensing Board of, 80, 94, 138, 154–155; Midland Plant and, 112, 114–115, 119, 128, 155, 191–192; Midland Plant settling and soils hearings by, 115, 117–118, 137, 139–140, 191; Midland Plant violations and, 147–148, 153, 156; NEPA and, 86; nuclear waste disposal and, 86–87; public affairs office of, 124, 152; quality control and, 112, 118–119, 124, 127, 147–148, 153; radiation regulatory changes by, 108; regulation changes and, 151, 172–173, 193; seismic design requirements of, 193; TMI2 radiation shut down and, 124–126; Vermont Yankee Case and, 219 (n. 37)

nuclear testing, atmospheric, 130

nuclear waste: disposal of, 87, 185; *Nuclear Power & Public Concern*, 56, 58, 60, 185–186. *See also* radioactive waste

Nute, Leslie, 102, 137

O

Oak Ridge Associated Universities, 66, 67–68

Oak Ridge National Laboratory, 23, 67–68, 107

Oakway IV, 143

Occupational Safety and Health Administration, 107

oil boom (MI), 29

oil prices, 96, 102

ORAU (Oak Ridge Associated Universities), 66, 67–68

Oreffice, Paul, 65–66, 138, 174

Osborn, John H., 28
Ott, Alan W., 70, 84, 92
Overberger, Charles, 12–13, 185
oxygen, 124, 188
Oyster Creek Plant, 7, 39–40

P

Palcich, Mary Jean. *See* Sinclair, Mary
Palen, John, 127
Palisades Plant: Bechtel bought interest in, 169; Bechtel contracted for, 41, 48, 101; Bechtel sued by, 41, 98, 101, 169; Campbell on, 40; Cherry and issues at, 70, 79–80, 82, 85, 217 (n. 8); Combustion Engineering reactor in, 41, 43; construction of, 40, 41, 48, 69, 72, 79; Consumers Power operated, 40, 52, 60, 69–70, 79, 101, 169; cost of, 41; interveners and, 69; location of, 40, 79; MW capacity of, 40–41, 43; off-line repairs and, 41, 96, 98, 169; permit for, 70, 72; PWR reactor at, 41, 43; sale of interest in, 169
Panhandle Eastern Corporation, 99, 175, 242 (n. 26)
Pardee, Elsa U., 216 (n. 6)
Parker, Robert (Bud), 92
Parker, Robert L., 70
Paton, William, 139
Peck, Bruce, 90, 91, 94, 104, 141, 152
Pediatrics (journal), 14
petroleum geologists, viii
Philadelphia Inquirer (newspaper), 93
Pile-1, 20, 187, 188
Pintek, Mike, 123–124, 228 (n. 6)
Planet of the Humans (Gibbs and Moore), 178, 243 (n. 42)
Plateau Resources Limited, 111
Pollard, William, 67
pollution: acid rain, viii, 37; from coal, 5, 30, 31, 36–38; controls, 37, 56; in Midland, 12, 30, 31, 53; nuclear, 12, 53, 56, 185, 227 (n. 22)
Poppe, Patricia Kessler, 178, 242–243 (n. 41)
Porter, William, 59, 212 (n. 23)
power plants: gas-fired, *176*. *See also* nuclear power plants
Power Reactor Development Co. v. Electricians, 16
Power Reactor Development Company, 15, 38, 161
PRDC (Power Reactor Development Company), Detroit Edison and, 15, 38, 161
President's Commission on the Accident at Three Mile Island (Kemeny), 127, 228 (n. 6)
protest marches: Michael Moore at, 134–134, *134*; in Midland, *130*, 133–135, *133*, *134*, *135*
Public Interest Research Group, 132
Public Service Company of Indiana, 172
PWR (pressurized water reactor), 23, 38, 41, 42, 43, 52, 188

Q

quality assurance (QA): AEC definition of, 92; cadwelding and, 92; Keeley as Consumers Power director of, 117; Metropolitan Edison's Three Mile Island plant and, 125; Midland Plant plagued by violations of, 92, 94, 112, 117, 147, 148, 153–155; noncompliance and, 94; NRC inspections for, 112; quality control included in, 92

"Quality Assurance at the Midland Nuclear Powerplant" hearing, 153, 235 (n. 8)
quality control (QC): AEC definition of, 92; Bechtel inspectors of, 149–151; cement and, 91, 93, 155; concrete and, 91; finances and, 118; foundations and, 117; Kemeny Report on, 127; at Midland Plant, 62–63, 89–94, 107, 112, 117–119, 147–149, 151–155; Midland "tendon sheaths" problems and, 112, 155, 225 (n. 6); NRC and, 107, 119, 127, 147; regulations and, 107; Selby and, 102, 119; violation fines and, 147–148; welding and, 149
Quality Improvement Policy (QIP), 152–153
Quanicassee nuclear plant: closing of, 97, 100, 162; Consumers Power and, 42; on Lake Huron, 42, 97; MW capacity of, 43, 178; Westinghouse PWRs at, 42, 43

R

Race, Patrick, 131–132, 139
racism, 78, 106, 182
radiation: atmospheric tests and, 58, 130; atomic bombs produce, 14, 187; background, 108; cancer from, 63, 125, 130, 186; Cohen on safety of, 108; dangers of, 13–15, 38, 58, 63; education workshops on, 68; effects of, 14; first in "The Atomic Plague," 14; ionizing, 187; leak at Fermi 1, 17, 187; leaks of, 60, 185–186, 244 (n. 1); levels of, 81; *Midland Daily News* on, 124; Midland Plant and, 130, 134–135; Midland Plant permit hearings and, 83; misinformation about, 67–68; *Nuclear Power & Public Concern* on, 58, 185; protests about, 134–135; Rickover on safety of, 24; "The Stunted Children of Hiroshima" on, 14. *See also* Three Mile Island Nuclear Plant
radioactive waste: Hanford tank leak and, 60, 185–186, 244 (n. 1); at Midland Plant, 185–186, 219 (n. 37); Sinclair on leakage of, 185; storage options for, 87, 186; Vermont Yankee case and, 85, 219 (n. 37)
Rajewski, James, 235 (n. 21)
Randall, Courtland, 67–68
Rasmussen Report (WASH-1400) (NRC), 67, 118–119, 214 (n. 11)
Rather, Dan, 147–148
reactor: breeder, 16, 17, 38, 47, 188; BWR, 38, 39, 41, 42, 43, 188; fusion, 60, 68; light-water, 38, 52, 188; PWR, 23, 188. *See also* nuclear reactors
regulatory challenges: AEC and, 17, 71; in Midland Plant, vii, viii, 71, 78, 84, 89, 107–108; NRC and, 17; NRC Midland Plant inspections and, 112; safety and, 67; Sinclair and, 132; and TMI2, 124. *See also* U.S. Nuclear Regulatory Commission
Rehnquist, William, 85
Reuther, Walter P., 15, 16, *16*, 17, 19
Reynolds v. Sims, 159
Rickover, Hyman: in AEC, 23; early life and education of, 21, 22–23; medal received by, *20*, 23, 24, 204 (n. 26); *Nautilus* (SSN-571) and, 19, 23; nuclear power knowledge of, 19–20, 23–25, 38; project management style of, 19–20, 24–25, 43, 94; Sawyer interview of, 21; Shippingport design and, 23, 24; U.S. Navy career of, 19–24, 25, 94; water design reactors and, 19

Rofan Energy, Inc., 174, 241 (n. 26)
Roger and Me (Moore), 132
Roodenko, Igal, 134
Roose, Kevin, 77
Rosengard, Pam, 131
Rudakas, David, 167
Rumple, Norman (Slim), 56, 57

S

Saari, Norm, 125, 164
Sabo, Wally, 152, 186
Sacramento Municipal Utility District, 52
Saginaw, MI, *2*, 28, 60, 73–74; Michael Moore and, 132
Saginaw Bay, 3, 60
Saginaw Intervenors, 60–62, 80, 83, 85–86, 87, 111–112
Saginaw River, *2*, 3; path of, 28, 60
Saginaw Township, 139
Saginaw Valley, *2*, 3; anti-nuclear events in, 60, 131; coal mines in, 5, 29; lumbering boom in, 5, 28; Native Americans of, 131, 238 (n. 7); newspapers in, 62. *See also* Bay City, MI; Flint, MI; Saginaw, MI
Saginaw Valley College, 60
Saginaw Valley Nuclear Study Group, 60, 68, 129–130; Cherry represented, 80, 81, 84–85
salts, 1–2
sand, 3, 90, 117–118
Saran Wrap, 30
Schakne, Robert, 147–148
Schuette, Bill, 176, *176*, 177
Scientific American (magazine), 58
Seiberling, John F., 153–154
seismic requirements, 193
Selby, John: on Bechtel's problems, 154; congressional testimony of, 153–154; as Consumers Power president, 42–43, 68, 95–102, 109, 119, 144, 157–158, 168–167; Dow contract and, 157; financial concerns of, 143–145, 157–158, 162; General Electric career of, 100–101; Kemeny Report and, 127–128; Midland Plant cancellation and, 162–165; Midland Plant opportunities for, 102, 145; quality assurance and, 153; quality control problems and, 107, 109, 127–128, 141; rate increases and, 158; retirement resignation of, 167–168
service water intake structure, 116, 117, 118
Shaw, Milton, 25
Shippingport, Pennsylvania, 7, 204 (n. 22)
Shippingport Atomic Power Station: civilian purpose example of, 20, 25, 39; commercial plant at, 20, 23, 39; costs of, 39; first fuel elements of, 23; intervenors, 24; kWh price of, 39; MW output of, 20, 24; PWR reactor at, 23, 24; Rickover oversaw development of, 23–24, 25; start-up of, 19–20, 24
"shit house" poet, 104
Sierra Club, 60, 62, 129
Silkwood (film), 231 (n. 26)
Silkwood, Karen, 134, 231 (n. 26)
Sims, Linda. *See* Lein, Linda
Sinclair, Ann, 62
Sinclair, Dick, 59
Sinclair, Mary: Asperger and, 82; assistant Sarah Hughes and, 130–131, 235 (n. 8); awards presented to, 55; *Bay City Times* story on, 62; as candidate for Michigan House of Representatives, 136, 165;

church of, 55, 58, 59; College of St. Catherine degree of, 57; description of, 55–57; dislike and shunning of, 57–59, 62, 65; on Dow Advisory Board, 62; on fusion development, 60; hired Cherry, 80; home and office of, *56*, 58, 62; as intervenor, 81, 131–132, 139, 213 (n. 29); in *Midland Daily News*, 56–57; Midland Plant delays and, 83, 87; Midland Plant opposed by, 59, 62–63, 121; Midland Plant protests and, 132; on Midland Plant safety, 53, 56, 69; MNPC opposed, 68–69; "Nuclear Environmentalism—A Case Study," 212 (n. 23); on *Nuclear Power, Benefits and Risks* TV program, 56; *Nuclear Power & Public Concern*, 56, 58, 60, 185–186; nuclear power safety concerns and activism of, 53, 56–60, 62–63, 121, 125, 130–131, 136, 186; on nuclear waste, 87; Palcich family of, 55; press reports on, 55; on radioactivity, 185; Saginaw Valley Nuclear Study Group and, 60, 68, 80, 129–131; soils hearings and, 139; support for, 59–60; teach-in events and, 60; Three Mile Island accident and, 125; *Toledo Journal of Great Lakes' Law, Science and Policy* article by, 63; UAW reprinted *Nuclear Power & Public Concern*, 58; at University of Michigan, 59, 60, 63, 212 (n. 23); U.S. Congress testimony of, 60; writing career of, 57–58

Sinclair, Peter, 57, 82

Sinclair, Rosemary, 57

Sinclair, William D., 57–58, 59

Sinderman, Roger, 125

Sixty Minutes (TV), 12, 201 (n. 14)

Skory, Gary, 212 (n. 21)

Smock, Mrs. Sidney N., 9, 68, 71

sodium (liquid coolant), 16, 19, 188

sodium bromide, 205 (n. 8)

sodium chloride (salt), 1

soil: clay, 3, 90, 116, 117, 192; fill dirt, 116, 117, 118, 192, *192*, 193; glacial till, 90, 116, 118, 192; gravel and, 90; sand, 3, 90, 117–118

solar generation, viii, 8, 43, 177; farm needs of, 178, 243 (n. 43); in Michigan, 178–179; MW capacity of, 243 (n. 43)

Source Cogeneration Company, 242 (n. 26)

Speak Up Now (organization), 8, 70

Speak Up Now rally (N-Rally): attendance at, 8, 73; bands at, 9, *9*; Detroit's Fat Bob sang at, 9, 201 (n. 11); Dow employees at, 8; Dow support and, 73; hostility in, 58–59; in Midland, 8–11, *9–11*, 12, 58, 70–71, 73; *Midland Daily News* on, 8, 12; Midland supporters of, 8, 10–12, 73, 78; MNPC financed, 8, 70; nuclear challenges at, 9–10, 12; speakers at, 9–10, *10*, 201 (n. 14); in Washington, D.C., 10, 11, *11*, 70–71

Special Project Labor Agreement, 106

Sporn, Philip, 8

stagflation, 97

Stamiris, Barbara, 139

steam: kp/hr as volume of, 177, 242 (n. 37); low-cost, vii

Steam and Electric Agreement (Dow and MCV), 175–176

steam generators, 188, 189, 195

steam turbines, 189, 192–193, *192*, 195, 241 (n. 15); in *The China Syndrome*, 122–123

steel, costs of, *108*

Stone & Webster, 155

Strahl, John, 152
Streep, Meryl, 231 (n. 26)
Strostacker, Charles J., 216 (n. 6)
styrene, 4, 58; polystyrene and, 182, 216 (n. 11)
Styrene: Its Polymers, Copolymers, and Derivatives (Boundy & Boyer), 58
submarines: *Nautilus* (SSN-571), 19, 23, *19*, *23*; nuclear, 19, 23, 25, 91, 151; traditional S-48, 22
sulfur dioxide, viii, 37
Swainson, John, 158

T

Taylor, Robert (Fat Bob the Singing Plumber), 9, 201 (n. 11)
teach-in, 60
Temple, Joseph G., 138
"tendon sheaths," 112, 155, 225 (n. 6)
Tennessee Valley Authority (utility), 25, 118, 172, 240 (n. 11)
Thornburgh, Dick, 125–126
Three Mile Island Nuclear Plant: accident at, 53, 87, 123–128, 135; design changes and, 53, 129, 140–141; emergency declared at, 123, 229 (n. 13); Harrisburg and, 134; Kemeny Report on, 127, 228 (n. 6); media and TV reports on, 123–124, 126, 128, 148, 234 (n. 3), 235 (n. 8); Metropolitan Edison operated, 52, 124–126, 127, 128; Midland Plant differs from, 127–128; Midland reaction to accident at, 124–127, 135–136; NRC and, 124; radiation from, 124–125, 128; Unit Two reactor (TMI2), 123, 126
Tittabawassee River: dam failures and floods on, 116, 182, 213 (n. 29); ecology and pollution of, 28, 139, 226–227 (n. 22); elevation of, 116, 227 (n. 26); floodplain of, 3–4, 5, 60, *61*, 115–116, 153, 182, 191; location of, *2*, 60, 81; Midland Plant site on, *6*, 60, *61*, *90*, 115; nineteenth-century condition of, 28; nuclear plants on, 5, 7, 60, 115, 134, 213 (n. 29); nuclear waste storage on, 186; thermal pollution in, 227 (n. 22)
Today (TV), 126
Todd, Charlie, 9, *9*
Toledo Journal of Great Lakes Law, Science and Policy, 63
"Treaty of Detroit," 16
Troffer, George, 125–126
turbine generator, 25
"turbine roll," 163–164
Turkey Point Four Plant, 48
Turkey Point Three Plant, 48
Turner, Arthur, 9
turnkey plant contracts, 40, 41
TVA (Tennessee Valley Authority), 25, 118, 172, 240 (n. 11)

U

UAW (United Auto Workers) (union): Fermi 1 safety and, 16–17, *16*; Local 599-Buick, 133; *Nuclear Power & Public Concern* reprinted by, 58; Reuther of, 15, *16*, 17
UCS (Union of Concerned Scientists), 129
Udall, Morris K., 153
Union of Concerned Scientists, 129
Union Local 85 (Pipefitters and Pipewelders), 105
unions: AFL-CIO, 71; apprentice programs in, 105, 106, 150, 166; International

Brotherhood of Electrical Workers, 150; International Union of Electrical Workers, 16; pipefitters, 105; power of, 17; UAW, 15–17, *16*, 58, 60, *133*; United Papermakers and Paperworkers, 16; welders, 105; women in, 105–106
United Auto Workers (UAW). *See* UAW
United Auto Workers International, 60
United Kingdom Atomic Energy Authority, 47
United Papermakers and Paperworkers, 16
United States Court of Appeals, District of Columbia (D.C. Circuit), 84
United States Testing, 154
University of Chicago, 20
University of Michigan, 150; Sinclair at, 59, 60, 63, 212 (n. 23)
University of Michigan Environmental Law Society, 60
University of Michigan–Flint, 132
University of Minnesota, 60
Uptegraff, Liz, 105–106, 224 (n. 10)
uranium: fissioning and, 4, 187; mining, 112; Plateau Resources Limited and, 111; prices, 112, 225 (n. 5); reactor core and, 186
uranium fuel (pellet), 4, 111, 114, 186, 187; byproducts, 86–87; price of, 112, 144, 225 (n. 5)
uranium oxide, 23
uranium 235, 4
U.S. Energy Research and Development Administration, 169, 214 (n. 6)
U.S. House of Representatives, 60, 84, 116, 148, 153
U.S. Naval Academy, 21, 22, 24, 28
U.S. Navy: napalm and, 216 (n. 12); Nuclear Power Branch, 23; Nuclear Power School, 91; Peck in, 90, 91, 94; Postgraduate School, 24; Rickover in, 21–24, 94; submarines in, 19, 22–23, 25, 91, 151
U.S. Nuclear Regulatory Commission (NRC), 83, 89, 107, 112, 124; on ACRS, 154; was AEC, 17, 214 (n. 6); Midland Plant problem solving by, 137, 147–148; on Midland wall cracking, 191; Rasmussen Report of, 67, 118–119; Region III inspection staff, 154; seismic requirements of, 193
U.S. Senate, 60, 84
U.S. Supreme Court: appeal of Midland permit Case in, 85–86, 87, 111–112; *Power Reactor Development Co. v. Electricians* in, 16; *Reynolds v. Sims* in, 159; Rehnquist opinion at, 85; *Vermont Yankee* case in, 85–86
utilities: accounting for profit by, 38, 207 (n. 15); economic stresses on, 100; investor-owned, 5, 17, 52, 145, 157, 161; MPSC regulated, 33, 142, 158, 161, 167, 171; nuclear plant cancellations by, viii, 100, 162–163, 165–166, 172–173; rate base return and, 161; rate increases by, 159–162, 171, 237

V

Valasek, Laurie, 242 (n. 37)
Van Dam, Tina S., 217 (n. 20)
Vermont Yankee Case (Natural Resources Defense Council v. U.S. Nuclear Regulatory Commission) (*Vermont Yankee Nuclear Power Corp. v. Natural*

Resources Defense Council, Inc., et al. and Consumers Power Company v. Aeschliman, et al.), 85, 219 (n. 37)
Vietnam War, 5; Consumer Price Index and, 51, 141; *Midland Daily News* articles on, 50–52; napalm in, 51
vinyl chloride, 4
Vogal plant, 106
Vrchick, Randy, 9, *9*

W

Wall, Harry, 35, 42
Wall Street Journal (newspaper), 80
War Resisters League, 134
Warren, Sharon, 139, 148
Watts Bar Unit 2, 172, 240 (n. 11)
Wessel, Milton, 71, 72
West Michigan Environmental Action Council, 60
Westinghouse: light-water PWR reactors of, 23–24, 38, 188; turnkey contracts and, 40
Wheeler, Bob, 94, 118, 119
White, George, *197*
Whiting, Helen Dow, 75, 216 (n. 6)
Whiting, Macauley, 29, 65, 71, 75, 216 (n. 6)
Whiting Forest, 182
Whiting Overlook Park, 4, 75; protest marches in, 133–134, *134*
Wick, George, 162
Wiecek, William M., 17, 202 (n. 5)
Wieland, Rodney, 115, 140
wind generation, viii; costs, 179; farms for, 177–178; Great Lakes, 178–179
Wood, James W., 14
Woodby, Pegg (Mudd), 107
Woodwell, George M., 58
World War I, 4, 73, 74
World War II, 4, 14, 73, 202 (n. 5); Bechtel in, 47; Rickover in, 23, 24
women: as Bechtel employees, 107; as Midland Plant employees, 104–107; as pipefitters and welders, 105–106, 224 (n. 10). *See also* Sinclair, Mary

Y

Yates, Dorothy Langdon, 27
Youngdahl, Russell, 41, 42, 71, 90, 107

Z

Zimmer Nuclear plant, 172
zirconium, 23, 114, 204 (n. 22)